SCIENCE

In Your World

SENIOR AUTHORS

Dr. Jay K. Hackett
Dr. Richard H. Moyer

Macmillan/McGraw-Hill
School Publishing Company

ACKNOWLEDGMENTS

For permission to reprint copyrighted material, grateful acknowledgment is made to the following authors, publishers, and agents. All possible care has been taken to trace the ownership of every selection included and to make full acknowledgment of its use. If any errors have inadvertently occurred, they will be corrected in subsequent editions, provided notification is sent to the publisher.

Harper & Row, Publishers, Inc.: Text from "Waves of the Sea" from *Out in the Dark and Daylight* by Aileen Fisher, illustrated by Gail Owens, text copyright © 1980 by Aileen Fisher. Illustrations Copyright © by Gail Owens.

William Collins and Sons: "Washing" by John Drinkwater, reprinted by permission of William Collins and Sons.

CREDITS

Series Editor: Jane Parker
Design Coordinator: Kip Frankenberry
Series Production Editor: Helen Mischka
Level Editors: David H. Mielke, H. Addison Lynes
Contributing Editors: Beth Britton Anderson, Beverlee Jobrack, Sharon Pruitt, Joyce R. Rhymer
Production Editor: Jillian C. Yerkey
Designer: Terry Anderson
Artist: Mark D. Clingan
Photo Editor: David T. Dennison

Macmillan/McGraw-Hill School Division
866 Third Avenue
New York, New York 10022

Printed in the United States of America

ISBN 0-675-16228-9

9 8 7 6 5 4 3 2

SENIOR AUTHORS

Dr. Jay K. Hackett
University of Northern Colorado

Dr. Richard H. Moyer
University of Michigan-Dearborn

CONTRIBUTING AUTHORS

Stephen C. Blume
Elementary Science Curriculum Specialist
St. Tammany Public School System
Slidell, Louisiana

Ralph M. Feather, Jr.
Teacher of Geology, Astronomy, and Earth Science
Derry Area School District
Derry, Pennsylvania

Edward Paul Ortleb
Science Supervisor
St. Louis Board of Education
St. Louis, Missouri

Dr. Barbara Swanson Thomson
Associate Professor in Science Education
The Ohio State University
Columbus, Ohio

CONTRIBUTING WRITER

Ann H. Sankey
Science Specialist
Educational Service District 121
Seattle, Washington

READING CONSULTANT

Barbara S. Pettegrew, Ph.D.
Director of the Reading/Study Center
Assistant Professor of Education
Otterbein College, Westerville, Ohio

SAFETY CONSULTANT

Gary E. Downs, Ed.D.
Professor
Iowa State University
Ames, Iowa

GIFTED AND MAINSTREAMED CONSULTANTS

George Fichter
Educational Consultant
Programs for Gifted
Ohio Department of Education
Worthington, Ohio

Timothy E. Heron, Ph.D.
Professor
Department of Human Services, Education
The Ohio State University
Columbus, Ohio

CONTENT CONSULTANTS

Robert T. Brown, M.D.
Assoc. Prof. Clinical
Pediatrics Dir., Section for
Adolescent Health The Ohio State Univ.
Children's Hosp. Columbus, Ohio

Henry D. Drew, Ph.D.
Chemist, U.S. FDA
Div. of Drug Analysis
St. Louis, Missouri

Judith L. Doyle, Ph.D.
Physics Teacher
Newark High School
Newark, Ohio

Todd F. Holzman, M.D.
Child Psychiatrist
Harvard Com. Health Plan
Wellesley, Massachusetts

Knut J. Norstog, Ph.D.
Research Associate
Fairchild Tropical Garden
Miami, Florida

James B. Phipps, Ph.D.
Prof., Geol./Oceanography
Grays Harbor College
Aberdeen, Washington

R. Robert Robbins, Ph.D.
Assoc. Professor
Astronomy Department
University of Texas
Austin, Texas

Sidney E. White, Ph.D.
Professor
Dept. of Geology/Mineralogy
The Ohio State Univ.
Columbus, Ohio

REVIEWERS: Teachers and Administrators

Annette Barzal, Walter Kidder Elementary School, Brunswick, OH; **Ronald Converse,** Conroe Independent School District, Conroe, TX; **Suzanne Doof,** C.E.S. 132, Fort Lee, NJ; **Lynda Frith,** Ball Intermediate, San Antonio, TX; **Paul W. Gates,** Madison Elementary, San Antonio, TX; **Shirley Gomez,** Luling Elementary School, Luling, LA; **Janice Gritton,** Gavin H. Cochran Elementary School, Louisville, KY; **Jhynelda Hahn,** Austin Elementary, Odessa, TX; **Glenn Hubert,** Miami Shores Elementary School, Miami FL; **Shirley Larges,** Azalea Middle School, St. Petersburg, FL; **Cynthia K. Leinweber,** Leon Valley Elementary, San Antonio, TX; **Janet McDonald,** Pine Middle School, Los Alamitos, CA; **Marsha McKinney,** Pope Elementary School, Arlington, TX; **Catherine Mitchell,** Shenandoah Elementary, San Antonio, TX; **Jeffrey Moniz,** St. Philomena Elementary School, Portsmouth, RI; **Sister Pauline Elizabeth Neelon,** St. Teresa Elementary School, Providence, RI; **Lynda Taylor,** Alta Vista Elementary School, Waco, TX; **Joy Tingle,** Terrebonne Parish School System, Houma, LA; **Jim Wilkerson,** Glenoaks Elementary, San Antonio, TX; **Jay Woodand,** Waukesha Public Schools, Waukesha, WI; **Eugene Wozniak,** Dearborn Schools Resource Center, Dearborn, MI

Table of Contents

Unit 1 Life Science 2

denotes ACTIVITY.
See page x for ACTIVITIES TABLE OF CONTENTS.

Unit 2 Earth Science 106

denotes ACTIVITY.
See page x for ACTIVITIES TABLE OF CONTENTS.

Unit 3 Physical Science 208

CHAPTER 10

CHAPTER 11

CHAPTER 12

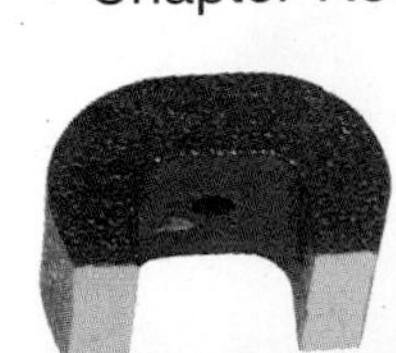

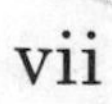

Unit 4 Human Body 272

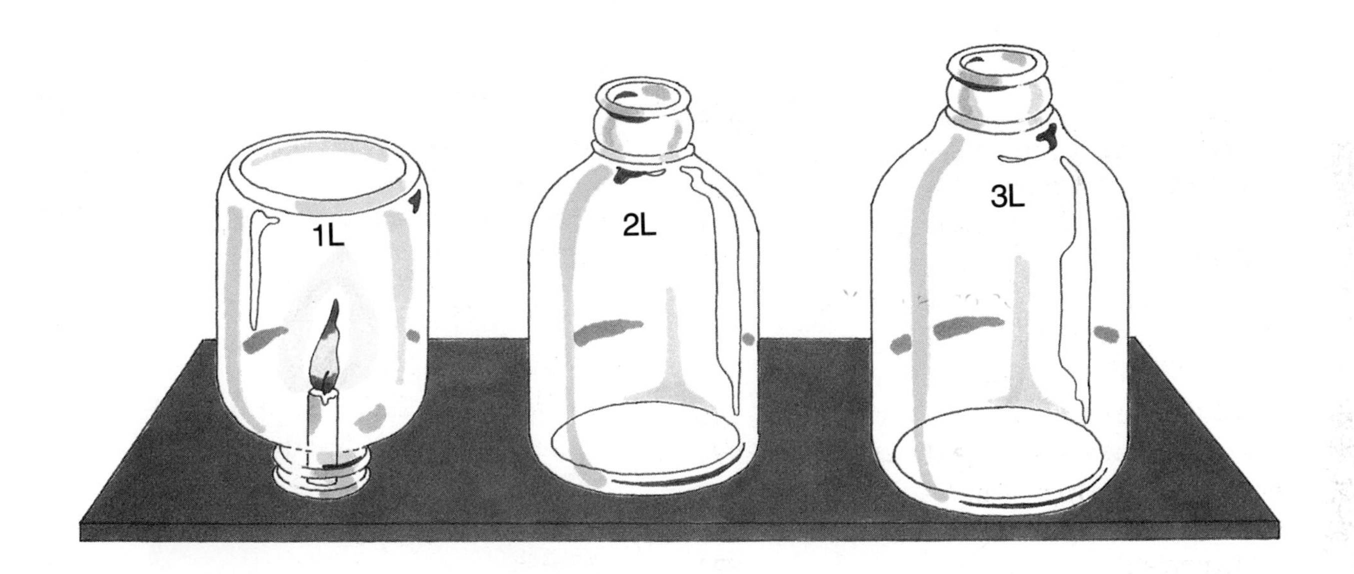

denotes ACTIVITY.

See page x for ACTIVITIES TABLE OF CONTENTS.

Activities

Have You Ever...

Activity

You Can...

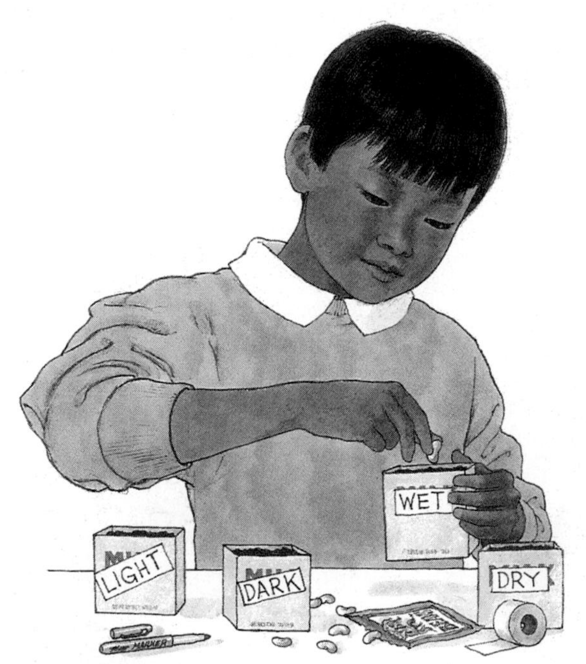

Process Skill Models

Problem Solving Activities

Science is . . .

Understanding

"Viewed from the distance of the moon, the astonishing thing about the earth . . . is that it is alive."

Lewis Thomas American physician (1974)

How does a sponge get its food?
Where do fish get oxygen to breathe?
How do plants make food?

Science has some answers for you.

Science is . . .

Discovering

"It is courage based on confidence,
not daring, and it is confidence
based on experience."
Jonas Salk
American physician (1955)

In 1955, Jonas Salk discovered a medicine that would keep people from getting polio, a crippling disease. Many people were afraid Salk's medicine would give them polio. Jonas Salk believed in his discovery, and he gave the medicine to himself, his wife, and his three sons.

Dare to discover science!

Science is... Deciding

Science is more than knowing.
When you use science, you make decisions
based on observations and experiments.
Science can help you decide.

Science is . . . Applying

"Having come so far, can we make the next invention in time?"
Margaret Mead
American anthropologist
(1901–1978)

Ask your grandparents to tell you what life was like before television. The idea for television came as early as the 1880s. By 1949, there were one million TV sets in America. Today 98% of the homes in America have at least one TV set. Did you watch TV today?

Science is...

Interpreting Data

Predicting

Hypothesizing

Find out all about these process skills on pages 343-353 of this science book.

in your world . . .

Life Science

UNIT 1

What do we plant when we plant the tree?
We plant the houses for you and me.
We plant the rafters, the shingles, the floors,
We plant the studding, the lath, the doors,
The beams and siding, all parts that be;
We plant the house when we plant the tree.

from "What Do We Plant?"
Henry Abbey

SCIENCE
In Your World

Scientific Method

Isaac Newton, one of the greatest scientists of all time, once observed an apple thudding to the ground. This made him think, "Why doesn't the moon fall also?" Observation and experimentation led Newton to scientific discoveries about gravity, movement of the planets, light, and color. You can be a junior scientist and explore your world using the same scientific method he used.

Have You Ever...

Wondered if Green Is Really Green?

Using a cotton swab, dab a tiny dot of green food coloring about 4 cm from one end of a strip of paper towel. Hang the strip over a glass of water so that the strip is in the water, but the spot is above the waterline. Let the strip stand overnight. How did the green dot change?

Importance of the Scientific Method

LESSON 1 GOALS
You will learn
- that observation is a way to study your world.
- that it's important to follow a plan when doing investigations.
- what the scientific method is.

Most people are curious about the world around them. They want to know what makes the weather change or why the leaves fall. They wonder why some plants grow better in warmer climates or why some animals prefer to live in cold climates. Look at the photo on this page. Suppose you saw these animals while you and your class were visiting the zoo. What kinds of things would you want to know about them, and how would you find out?

There are different ways to find answers to your questions. You can use your senses—feel, smell, hear, see, taste—to find out more about some things. But for others, you have to follow a careful plan to find the information you want.

An aardvark and a platypus

Observation is one way to study your world. However, if you don't write down details about what you observe, you can't share the information you learn with others. If you don't record the steps you used to do an activity, you may not be able to tell someone else how to repeat them.

What is one way to study your world?

Let's look at the ways Jack and Carol used to find the answers to their questions.

Carol and the Geraniums

Carol wanted to find out whether geranium plants would grow better in sunlight or in shade. She thought the plants would grow better in sunlight. To begin her investigation, Carol found two geranium plants that were the same size. She placed one potted plant in a window that gets a lot of sun and the other in a shaded part of the house. She tried to remember to water the plants regularly, but sometimes she forgot. At other times, she gave more water to the shaded plant than to the one in the window.

Carol watering her plants

Carol was surprised to find that after two weeks, the shaded geranium had grown taller than the geranium in the sunny window. Carol was puzzled. What went wrong with the investigation? Why was the shaded geranium taller?

Jack and His Gerbil

Why did Jack think his gerbil would prefer red food?

Jack had a pet gerbil at home. He knew that some scientists do experiments to find out if animals prefer food of a certain color. Jack thought his gerbil would prefer food that is colored red because he had seen it gnaw on a red block of wood in the cage. Jack decided to try this experiment. He took 16 sunflower seeds and used food coloring to color four seeds yellow, four seeds red, and four seeds blue. He left four seeds uncolored. Jack placed the colored seeds in four containers of the same type. He put the containers where the gerbil could reach each of them easily. He was careful to make sure all other conditions remained the same inside the gerbil's cage.

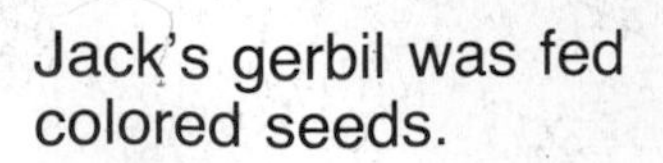

Jack's gerbil was fed colored seeds.

What differences did you notice in the way Jack and Carol made their observations? Which method would probably produce the most useful information?

Scientists study bird migration.

Scientists Ask Questions

Why is it important to use an orderly plan when looking for answers to questions? One reason that scientists use scientific methods when they investigate questions or problems is so they can repeat the experiment many times. By repeating experiments, scientists can be more sure that the answers they discover are correct. Also, sometimes other scientists want to do the same experiment. They can follow the steps in the experiment in the same order.

As part of their jobs, scientists ask questions that help them find answers. They collect information in organized ways to help them in their search for possible answers. We call this organized way to gather information the **scientific method.** There are five steps in the scientific method you will use this year, and you will study them in the next lesson.

ACTIVITY

You Can...

Be a Plant Scientist

Plant scientists, or botanists, observe how plants grow. You can be a botanist. Find a plant with a straight stem. Put a ruler straight up in the soil next to the stem. Place the plant by a sunny window. Observe the leaves and stem for several days. What happened to the leaves? What happened to the stem? Explain why this happened.

Lesson Summary

- You use your senses to make observations of your world.
- It is important to use orderly steps when trying to find answers to questions.
- The scientific method is an organized way of asking questions, gathering information, and finding answers.

Lesson Review

1. How do you make daily observations?
2. Why do scientists use scientific methods?

★3. Compare the experiments Carol and Jack did. Who followed orderly steps? How could the other experiment have been improved?

How does gravity affect plant growth?

What you need

2 baby food jars with lids
(labeled A and B)
8 radish seeds
paper towels
water
spoon
pencil and paper

What to do

1. Line the inside of each jar with a paper towel.
2. Use the spoon to add equal amounts of water to each jar until the paper towels are moistened.
3. Put 4 radish seeds between the paper towel and the side of each jar.
4. Put on the lids, and set the jars in a bright place.
5. Record your observations each day. You will observe the seeds for a total of 5 days.
6. After 3 days, place jar B on its side.
7. Write a hypothesis about how gravity might affect growth.

What did you learn?

1. How does gravity affect plant growth?
2. What factors were kept the same for jars A and B?
3. What factor did you change for jars A and B?
4. Why didn't you turn jar A on its side?

Using what you learned

1. Can you explain why trees on a hill grow straight up?
2. How are you like a scientist?

Steps of the Scientific Method

LESSON 2 GOALS
You will learn
- the steps of the scientific method.
- that scientists need a common language.
- to use the steps of the scientific method.

Carlos observed that the plants by his classroom window grew tall and looked healthy. He also noticed that the plants in the darker part of the classroom looked different. The window plants had ten hours of bright light every day and looked much better than the plants that had less light. Carlos wondered if 24 hours of light would make healthier and taller plants. He decided to do an experiment to find an answer to his question. He chose bean plants for his experiment.

Carlos used a five-step plan in looking at his plants. Scientists often follow these same steps. You can use the same method to learn more about your world.

1. State the Problem (Ask a Question)

First, Carlos made an observation about the plants in his classroom. Plants near light grow tall and look healthy. Then, he asked a question. Would bean plants grow taller and look healthier with 24 hours of light? This is the question Carlos wanted to answer.

Plants by the window

2. Form a Hypothesis

Carlos made some observations about the classroom plants and suggested an answer to his question. His answer or **hypothesis** is a statement that tells what he expects will happen in his experiment. Carlos' hypothesis is:

What is a hypothesis?

Plants that get 24 hours of light per day will grow taller and look healthier than plants that get ten hours of light per day.

Carlos writes his hypothesis.

3. Design the Experiment

Carlos needed to decide how to test his hypothesis. First, he made a list of steps he needed to follow. Next, he made a list of equipment he needed to do the experiment. Next, he would treat the plants the same except for the amount of light. He decided that one group of plants would receive ten hours of light per day. The other group would receive 24 hours of light each day.

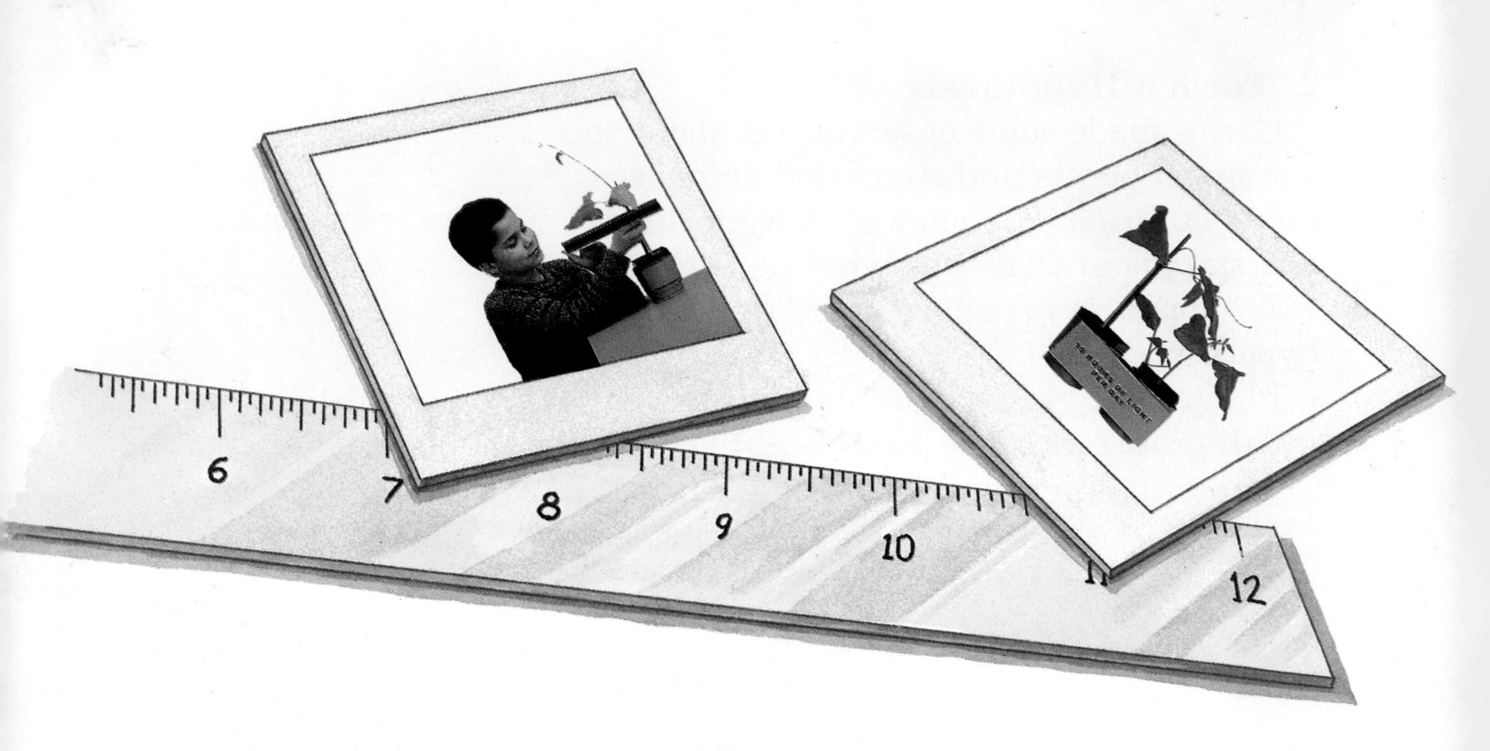

4. Record and Analyze the Data

What are data?

Carlos collected the data by taking measurements and describing his observations in a journal. He measured the bean plants and recorded his information on a chart as data. **Data** are recorded facts or measurements from an experiment. Then he studied the data in his journal and on his chart to see what had happened to the plants.

5. Draw a Conclusion

When Carlos ended his experiment, he needed to draw a conclusion. He knew this **conclusion** would be the answer to his question. Based on his study of the data, Carlos concluded that ten hours of light produced taller, healthier bean plants. Then, Carlos asked himself two questions.

What did I find out? (his conclusion)
Is this what I thought would happen? (his hypothesis)

When he compared his conclusion to his hypothesis, he discovered new information. He found that his hypothesis was false. His hypothesis stated that bean plants with 24 hours of light per day will be taller and healthier. However, from his experiment he found that ten hours of light per day was better for the plants.

Carlos can discuss his experiment with other people easily because he used a method to conduct his experiment. He used the same terms as scientists use. Carlos can talk about his observations, hypothesis, data, and conclusions.

Carlos enjoyed using the scientific method. He would like to do this experiment again. Next time, however, instead of observing four plants for ten days, he wants to observe 20 plants for 15 days. Do you think the outcome will be the same? How could he state the problem? What hypothesis could he use? How should he organize the data?

Many scientists retest other scientists' hypotheses. You might want to retest Carlos' hypothesis. Or you can use your own observations to state a problem, form a hypothesis, design, and carry out an experiment using the scientific method.

Table 1 The Steps of the Scientific Method
1. **State the Problem** Make observation and ask a question.
2. **Form a Hypothesis** Make your best guess about the answer to your question.
3. **Design the Experiment** Decide how you will explore your question.
4. **Record and Analyze the Data** List your observations and measurements. This is your data. Study the data to find out what has happened in your experiment.
5. **Draw a Conclusion** Based on your study, decide what the data mean. The conclusion will be an answer to your question.

Others can test a hypothesis.

Lesson Summary

- The five steps of the scientific method are: state the problem, form a hypothesis, design an experiment, record and analyze data, and draw a conclusion.
- The detailed steps of the scientific method make it easier for us to share information with other people.
- You can use the scientific method to explore the world around you.

Lesson Review

1. What are the steps of the scientific method?
2. What was Carlos' question?
3. What conclusion did Carlos draw? How did it compare with his hypothesis?

★4. If you were going to repeat Carlos' experiment, how would you set it up?

Application of Scientific Method to Life Science

LESSON 3 GOALS
You will learn
- how a scientist used the scientific method in her work.
- how the scientific method provided valuable information about saving the mountain gorillas.

In the previous lessons in this chapter, you have learned how to use the scientific method to explore your world. You may have made some discoveries that have given you the answers to questions you've wondered about. By now, you probably realize that the scientific method is a valuable tool for solving problems.

One big problem that scientists are trying to solve has to do with animals. Some animals are having a hard time surviving in their habitats. One of these animals is the gorilla. Scientists are trying to find out why so many gorillas are dying. These scientists often go to live near the gorillas they are trying to save.

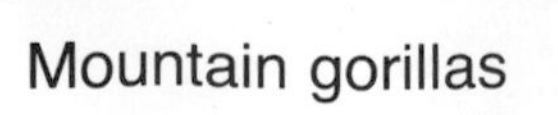

Mountain gorillas

Dian Fossey was especially concerned about the mountain gorillas in Africa. They were disappearing from places where they had always lived, and she wanted to know why. Did they have enough food? How much space did a gorilla need to live? Fossey realized that she needed to see the problem closeup before she could find a solution. In 1967, she went to live in the Virunga Mountains in Africa to study the mountain gorillas.

The Virunga Mountains

Dian Fossey needed to answer the question "What causes the number of mountain gorillas to keep decreasing?" Each year there were fewer gorillas. Scientists predicted that within ten years, there might be no more mountain gorillas in the world. The Virunga Mountains had all that were left. To help answer this question, Fossey hypothesized that loss of habitat was an important reason for the decrease in gorillas.

What question did Dian Fossey need to answer?

Dian Fossey taking notes

Observing Gorilla Behavior

It's very hard to study animals in their natural habitat. Wild animals are usually afraid of people and will avoid them. First, Fossey had to find the gorillas. They lived in forests located on steep mountains. Other scientists had not been able to get close to them. She knew she would have to get close to the gorillas to study them and collect accurate data. For months and months Fossey followed the gorillas. She could always hear them in the distance, but she only saw them for a moment now and then. They ran away whenever they saw her. But she was very patient, and finally the gorillas let her watch them from a distance.

What gorilla behaviors did Dian Fossey observe?

As she watched, Fossey observed some interesting gorilla behaviors. She observed that gorillas use special sounds to talk to each other. She observed that they make other sounds when they are eating. She became good at imitating these sounds.

She also imitated other gorilla behaviors. Because gorillas are knuckle walkers, they are not very tall when they walk. Fossey was very tall, so she bent over when she walked to make herself look shorter. Gorillas groom themselves to keep their fur neat and tidy. Fossey pretended to groom herself, too.

Slowly the gorillas learned to trust her and allowed her to come closer. She became their friend and was able to learn many things about them. One surprising thing she learned was that these large animals are gentle giants. She made careful observations and wrote down all of her data in her journals.

What surprising thing did Dian Fossey learn about the gorillas?

ACTIVITY

You Can...

Be a Wildlife Detective

Be a good observer and see what kinds of animals live in your neighborhood. Some animals look for food very early each morning or late in the day. Sometimes it's hard to find these animals. Be very quiet and very observant. How many different kinds of animals live near you? Use books in your library to classify these animals. Make a list. Did you discover some animal neighbors you did not know you had?

SCIENCE AND . . .
Math

The blue whale, an endangered animal, can weigh 150 metric tons. This equals the weight of 4,286 fourth graders. What is 4,286 rounded to the nearest hundred?

A. 4,300
B. 4,390
C. 4,200
D. 4,290

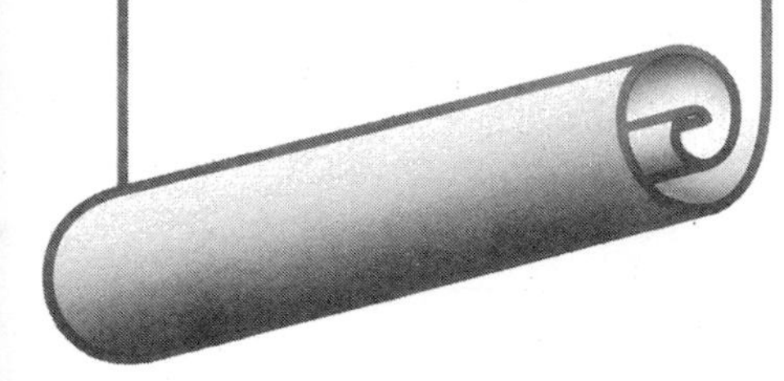

What were Dian Fossey's conclusions?

As Fossey analyzed the data in her journals, she discovered many interesting patterns. She found that the gorilla habitat was being taken away by farmers. They cut down the trees and planted crops. Loss of habitat was a big problem. She discovered that an even bigger problem was that people were setting illegal traps to catch deer and other animals. Many gorillas died because they were caught in these traps.

After Fossey analyzed her data, she drew some conclusions. She concluded that there were fewer gorillas each year because 1) they were losing their habitat to farmers, and 2) they were being caught in traps. These conclusions helped her to make a plan for saving the remaining mountain gorillas.

Dian Fossey shared her scientific study with the world. Laws were passed. Now farmers can't cut down the forests where the gorillas live. Traps are being collected and destroyed. People now realize how important it is to save these animals.

A gorilla trap

Without Dian Fossey there probably wouldn't be any mountain gorillas alive today. She died in 1987, but other scientists are continuing her work in the Virunga Mountains in Africa. They continue to follow her plan to save the mountain gorillas.

Fossey's work has helped protect mountain gorillas.

Lesson Summary

- Some scientists go to where animals live to study them. They use scientific methods to study these animals. Observations are an important part of a scientific method. Scientists use careful observations to collect data.
- Dian Fossey used careful methods of observation and recorded data about mountain gorillas. Her work helped convince others to work at saving them.

Lesson Review

1. Why did Dian Fossey go to Africa?
2. What steps did she follow in her work?

★3. Suppose a dog followed you to school. How might you use the scientific method to locate the dog's home?

How can data be collected?

What you need

15 fresh green beans
metric ruler
paper towels
pencil and paper

What to do

1. Use the ruler to measure the length of each bean. Round the measurement to the nearest centimeter.
2. Place a slash mark (/) under the length of each bean on the data table.
3. When all the beans have been measured, count the number of slash marks and write this number as the total number of beans.
4. Make a bar graph of the total number of beans at each length. Label the bottom "Length of beans in centimeters" and the side "Number of beans."

What did you learn?

1. What was the shortest bean you measured? The longest?
2. Which length of bean occurred most often?

Using what you learned

1. How can you collect and organize data?
2. Why do scientists often use tables and graphs?
3. What other characteristics might be different in your group of beans?
4. How would you collect data to show the number of students having birthdays in different months?

I WANT TO KNOW ABOUT...

Outlining

An outline is a list of important facts about a topic. It is a type of summary.

(1) The title of an outline names the subject.

(2) Each topic in an outline follows a Roman numeral and a period.

(3) Each subtopic in an outline follows a capital letter and a period. Here is an example of an outline. What are the parts?

The Camping Trip

I. Morning
  A. Got dressed
  B. Ate breakfast
  C. Gathered camping gear
  D. Left in the car

II. Afternoon
  A. Arrived and set up camp
  B. Went on a hike
  C. Fixed and ate dinner
  D. Built a campfire

Read the following paragraph and make an outline.

Jellyfish

Jellyfish are animals that look like open umbrellas. Some jellyfish are brightly colored, while others may be a milky shade. They may be as small as your thumb or as big as a grown person. Lying on a sandy beach in the sun, a jellyfish may not even look like an animal.

CHAPTER REVIEW 1

Summary

Lesson 1

- You use your senses to make observations of your world.
- It is important to use orderly steps when trying to find answers to questions.
- The scientific method is an organized way of asking questions, gathering information, and finding answers.

Lesson 2

- The five steps of the scientific method are state the problem, form a hypothesis, design an experiment, record and analyze data, and draw a conclusion.
- The detailed steps of the scientific method make it easier for us to share information with other people.
- You use the scientific method to explore the world around you.

Lesson 3

- Some scientists go to where animals live to study them. They use scientific methods. Observations are carefully made in order to collect data about the animals.
- Dian Fossey observed and recorded data about mountain gorillas. She helped convince others to save the gorillas.

Science Words

Fill in the blank with the correct word or words from the list.

scientific method **data**
conclusion **hypothesis**

1. A(n) ___ is a suggested answer to a question.
2. A(n) ___ is an actual answer to a question.
3. Measurements from an experiment are ___.
4. An organized way to gather information is the ___.

data

Questions

Recalling Ideas

Correctly complete each of the following sentences.

1. There are five steps in the
   (a) hypothesis.
   (b) conclusion.
   (c) scientific method.
   (d) observations.
2. Dian Fossey began her study of the mountain gorillas by first
   (a) stating the problem.
   (b) forming a hypothesis.
   (c) designing an experiment.
   (d) recording data.
3. One way to study your world is to
   (a) use your senses.
   (b) make observations.
   (c) use the scientific method.
   (d) all of these

Understanding Ideas

Answer the following questions using complete sentences.

1. Why should you write down details about observations?
2. Why do scientists repeat experiments many times?
3. What are the steps of the scientific method?
4. What is the advantage of using the scientific method and scientific terms when conducting an experiment?
5. How did Dian Fossey's investigation eventually help the mountain gorillas?
6. What conclusions did Dian Fossey draw after analyzing her data?

Thinking Critically

Think about what you have learned in this chapter. Answer the following questions using complete sentences.

1. What is the importance of using the scientific method in all life science investigations?
2. If your pet dog began sleeping much more than usual, how would you go about determining why he is doing so?

SCIENCE
In Your World

Invertebrates

You may have seen a colander being used to strain spaghetti. As water flows through the colander, the spaghetti is trapped in it. The colander acts as a filter.

Mollusks, like cockles, mussels, oysters, and clams, have filters, too. Their filters are used to sift food from water drawn in. No wonder they are called filter feeders.

ACTIVITY

Have You Ever...

Filtered Dirty Water?

Mix bits of soil, twigs, and plant material in a cup of water. Then pour the dirty water mixture through a coffee filter into a bowl. Examine the water in the bowl. How is the water in the bowl different from the original mixture? Examine the coffee filter. What does a filter do?

Simple Invertebrates

LESSON 1 GOALS
You will learn
- why some animals are called invertebrates.
- some characteristics of simple invertebrates.

Have you ever visited a zoo? Think about all the different animals you saw there. You have probably thought about how some are alike or different. What you may not have realized is that all animals can be arranged into two groups—animals that have backbones and animals that don't. Animals that don't have backbones are called **invertebrates.**

Sponges

The simplest kind of invertebrate is a **sponge.** This sponge is not like the pink or blue sponges you probably have around your house. Living sponges are usually found on rocks or other objects on the ocean floor. They don't move around. Their bodies don't have definite shapes, but they have many holes, or pores, that carry water throughout their bodies. Water flows in through the pores and out through the top of the sponge. The water contains food and oxygen for the cells of the sponge.

A variety of sponges

ACTIVITY

You Can...

Be an Interviewer

Sometimes scientists interview people to get data. You can be an interviewer, too. Interview 15 friends and neighbors about the use of sponges. Record their answers to these questions. 1. Have you ever used a natural sponge? 2. Do you use natural or artificial sponges now? 3. Where do natural sponges come from?

If part of a sponge is damaged, the cells of the sponge will regrow a whole new sponge. This regrowth is called **regeneration.** Regeneration of cells occurs in all animals. However, animals with simple body parts, like sponges, are able to regenerate major body parts.

Stinging-celled Animals

If you have ever been to the ocean, you may have seen a jellyfish that washed up on the beach. People who live by the ocean know enough about jellyfish to leave them alone when they see them. Jellyfish have stinging cells that cause very painful stings.

Stinging-celled animals

Jellyfish, hydra, sea anemones (uh NEM uh neez), and coral all belong to the group of animals with stinging cells. **Stinging-celled animals** live in water. They are simple invertebrates that have hollow bodies with only one opening—a mouth. The mouth is surrounded by tentacles (TENT ih kulz) with stinging cells that catch food and pull it toward the mouth. They then stuff the food into a body tube where it gets digested.

Flatworms

Flatworms are simple worms with long, soft bodies that are flat. Their solid bodies have only one opening for taking in food and getting rid of wastes. Flatworms have cells that form tissues for sensing their surroundings. Flatworms are more complex than sponges or stinging-celled animals. They have organs for moving, digesting food, and getting rid of materials. A flatworm is another animal that undergoes regeneration. If part of a flatworm is damaged, cells will regrow the part.

How are flatworms more complex than sponges or stinging-celled animals?

A flatworm may live as a parasite inside the body of another animal. A **parasite** is an organism that uses another animal to supply it with food, water, and oxygen. Tapeworms are flatworms that are parasites in the bodies of animals such as cows, pigs, fish, and humans.

What is a parasite?

Some flatworms are scavengers (SKAV un jurz). **Scavengers** feed on dead or rotting organisms. A planarian (pluh NER ee un) is a scavenger that swims and feeds on dead animals and plants in water.

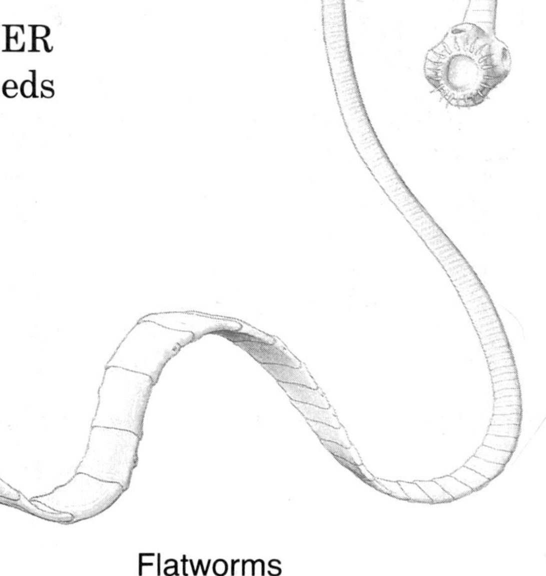

Flatworms

Roundworms

Roundworms have bodies that are more complex than the bodies of flatworms. There are two openings instead of one. Food enters through one opening, and wastes leave by the other. Many roundworms are parasites. One type of roundworm lives inside pigs.

Remember, even though the animals in this lesson may not all look alike and may live in different places, they are alike in some important ways. Can you name some of these ways?

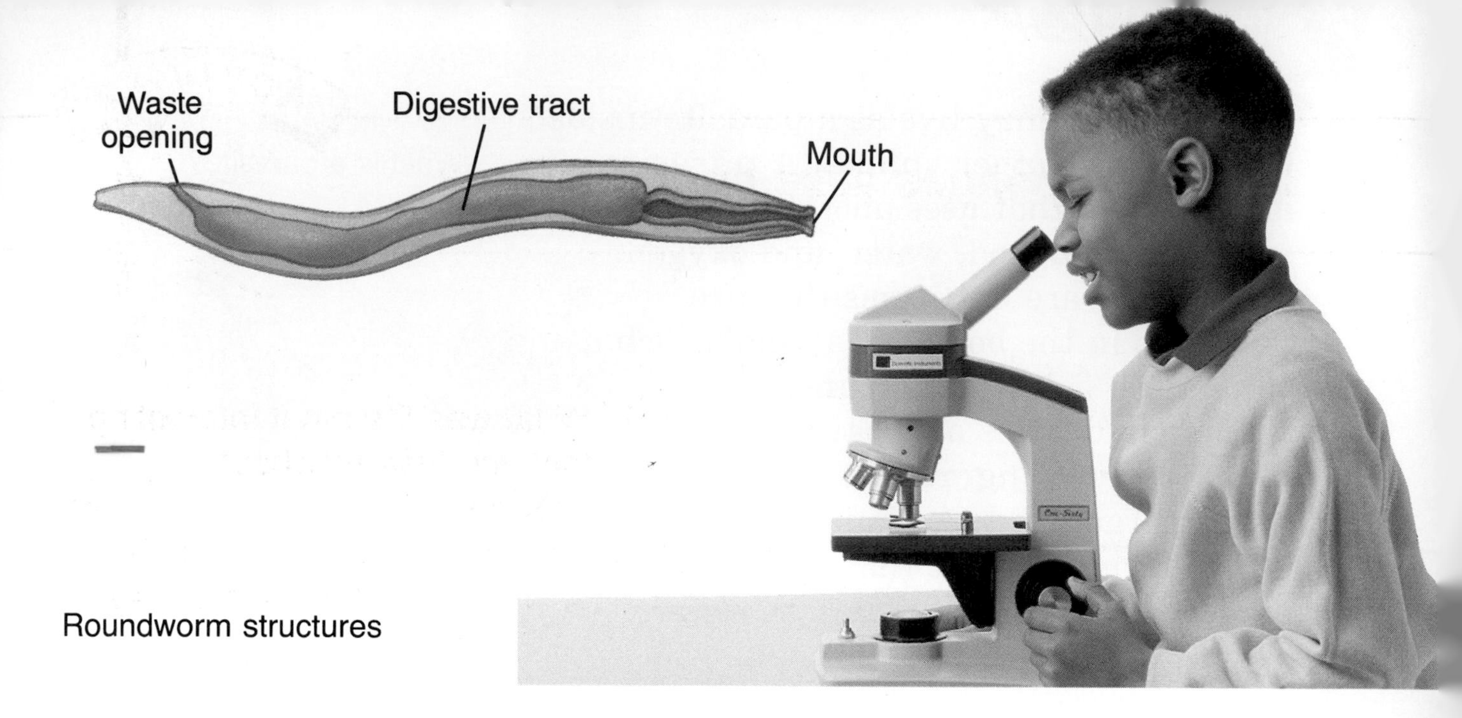

Roundworm structures

Lesson Summary

- Invertebrates are animals that don't have backbones.
- Sponges are simple invertebrates with many pores through which water flows. Food and oxygen in the water are used by the cells of the sponge. Stinging-celled animals have tentacles that are used for capturing food. Flatworms and roundworms have tissues and organs for moving, digesting food, and getting rid of waste materials.

Lesson Review

1. Name the four invertebrate groups from simplest to most complex.
2. What is regeneration? Name two animals that can regenerate body parts.
3. How do stinging-celled animals obtain food?
★**4.** How are parasites different from scavengers?

How are sponges alike and different?

What you need

natural sponge
artificial sponge
hand lens
reference materials
pencil and paper

What to do

1. Handle the natural sponge. Use a hand lens to observe this sponge. Record any observations.
2. Repeat step 1 with the artificial sponge.
3. Compare the natural with the artificial sponge. Record your observations.
4. Use reference materials to answer these questions.
   (a) Is a sponge a plant or animal?
   (b) Does it have roots?
   (c) Does it make food?
   (d) What are the pores?
   (e) How does it move?
   (f) What happens if a sponge is split into two parts?
5. Pretend you are a sponge. Keep a diary of your life for five days.

What did you learn?

1. What was the most interesting fact you learned about sponges?
2. How are natural and artificial sponges alike? How are they different?

Using what you learned

1. What characteristic of natural sponges is useful for survival?
2. Which type of sponge is probably more expensive to buy? Why?

Complex Invertebrates

LESSON 2 GOALS
You will learn
- the difference between simple and complex invertebrates.
- some characteristics of complex invertebrates.

In the first lesson of this chapter, you learned that all animals can be arranged in two groups—those that have backbones and those that don't. Do you remember that animals without backbones are called invertebrates? Some of the animals in this lesson you might find in your garden or even on your dinner plate. These animals are also invertebrates, but they're more complex than the ones you just studied.

Complex Invertebrates

How are complex invertebrates different from simple invertebrates?

Complex invertebrates are different from simple invertebrates because they have a larger number of cells that form tissues. These tissues, in turn, are grouped to form organ systems. Complex invertebrates have circulatory systems and nervous systems. Mollusks, segmented worms, arthropods, and spiny-bodied animals are complex invertebrates. These organisms all have a body tube where food is digested. This tube has an opening at each end.

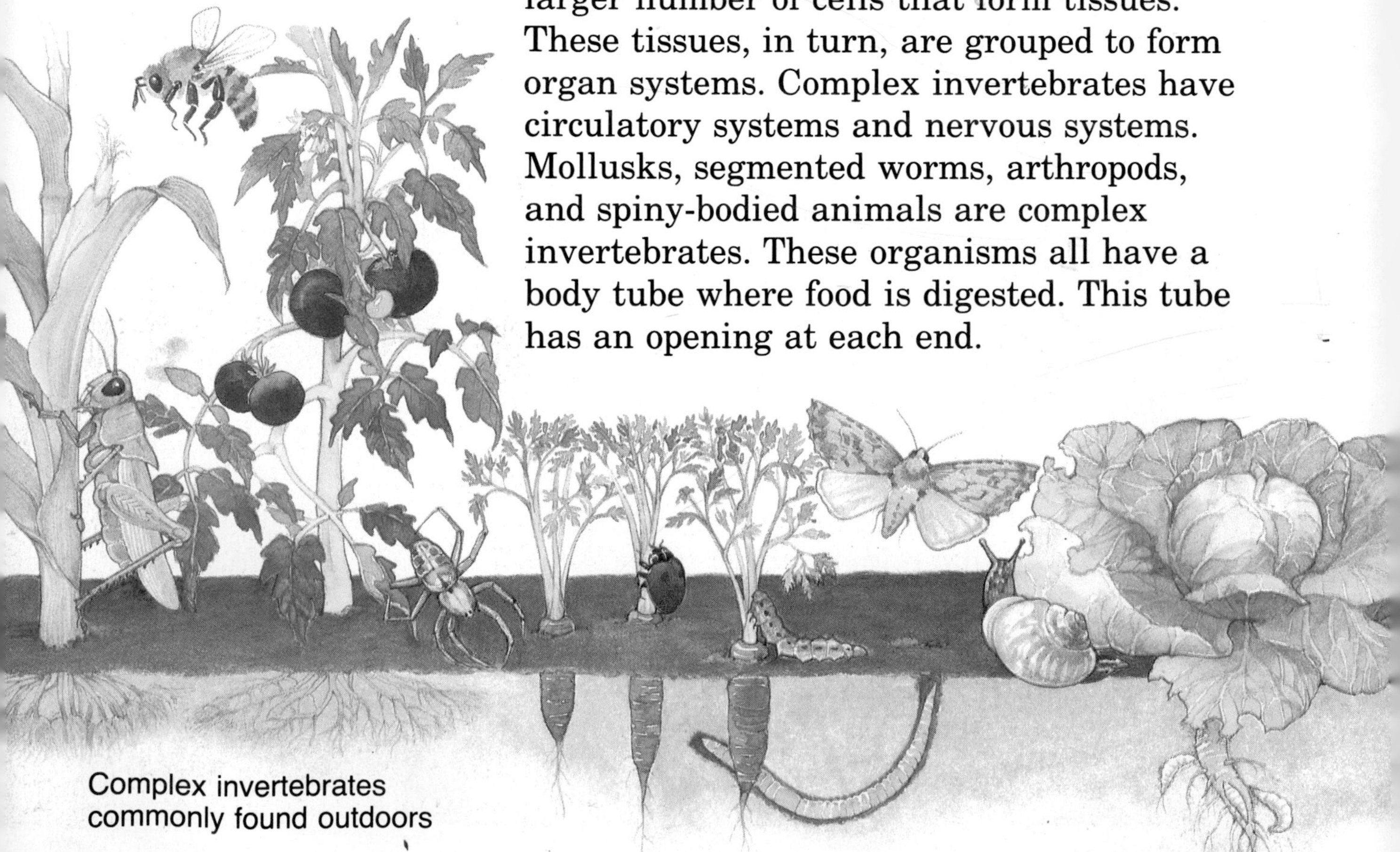

Complex invertebrates commonly found outdoors

Mollusks

Can you think of a way that a snail and an octopus are alike? They don't look alike do they? However, both of them are mollusks. Mollusks are a very large group of animals. They are grouped by their characteristics.

Mollusks are soft-bodied invertebrates that live on land or in salt or fresh water. Many mollusks have shells that protect their soft bodies. Snails have one shell. Clams and oysters have two shells with a hinge. Other mollusks, such as octopuses, don't have shells. Most mollusks have a head and a muscular foot for moving. Did you know that people use all these mollusks for food? How would you like to have a nice dish of snails for lunch?

Some mollusks don't have outer shells.

Segmented Worms

How are all segmented worms alike?

Segmented worms are also complex invertebrates. They are complex because they have organ systems for digesting food, getting rid of wastes, and moving blood. Earthworms, leeches, and clamworms belong to this group. They all have bodies that are divided into rings, or segments. Some segmented worms live on land, and others live in salt or fresh water.

Segmented worms live in different environments.

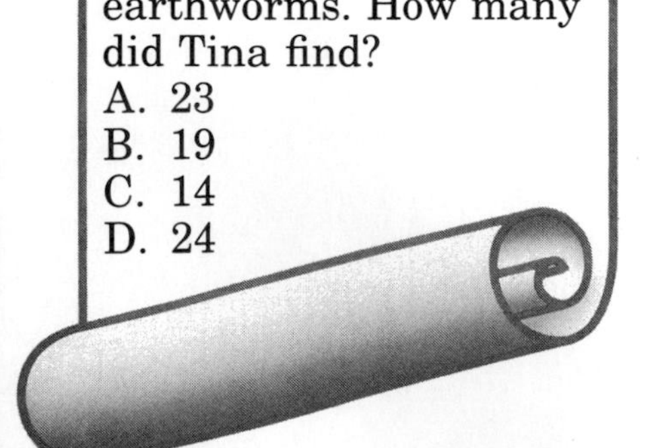

SCIENCE AND . . .
Math

Tina's science teacher asked her to collect some invertebrates. She collected 11 planaria, 8 slugs, and 5 earthworms. How many did Tina find?

A. 23
B. 19
C. 14
D. 24

Arthropods

Arthropods are invertebrates that are more complex than mollusks or segmented worms. An **arthropod** has an outer skeleton, legs that bend at joints, and a body divided into sections. Each body section has body parts that do special jobs. Arthropods have sense organs for seeing, feeling, tasting, and smelling, as well as organs for breathing. An arthropod's outer skeleton, called an exoskeleton, protects the organs inside its body. An arthropod must molt or shed its exoskeleton as it grows because the exoskeleton doesn't grow with the animal.

Arthropods are the largest group of animals. Because of this, they are divided into smaller groups made up of animals with similar characteristics. Centipedes and millipedes are two groups of arthropods that look like segmented worms with legs. They have many pairs of jointed legs, a head, and many segments. These animals live on land, often under stones or logs. They usually feed at night and hide during the day.

Crabs, crayfish, shrimp, and lobsters are part of another group of arthropods. These animals have two body sections and jointed legs. Most of the animals in this group live in water, but some live in damp places on land.

Insects are the largest group of arthropods and live almost everywhere on Earth. **Insects** have three body sections—a head, a middle section with three segments and three pairs of legs, and a stomach. Many insects also have wings. Some insects are helpful to people, and some are harmful. Can you think of some insects that help you? Can you think of some that you like to avoid?

Would You Believe?

Crickets have sense organs for hearing on their front legs instead of on their heads.

What is the largest group of arthropods?

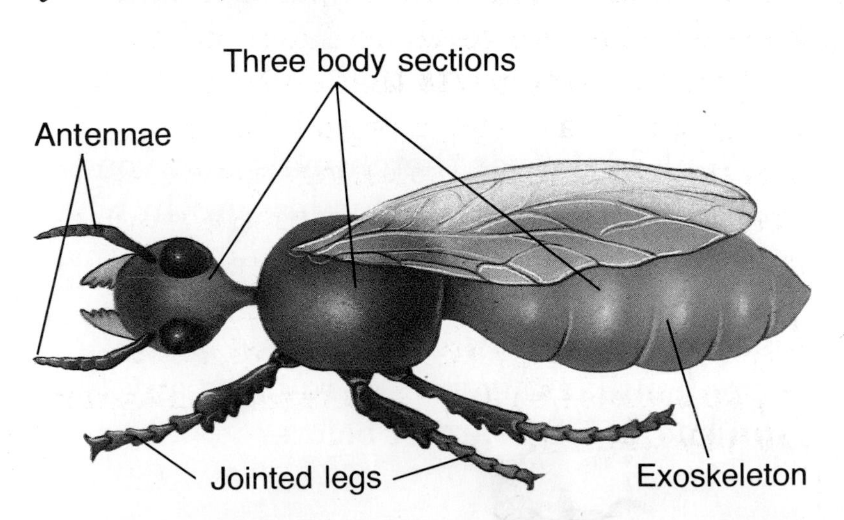

Insect body parts

Spiders, ticks, mites, and scorpions make up another group of arthropods. These animals have four pairs of legs and two body sections. Most of these animals live on land.

Spiny-bodied Animals

Starfish, sea urchins, sand dollars, and sea lilies make up another group of invertebrates. These animals have five-part bodies and spines. The body wall of these animals contains a skeleton of plates. **Spiny-bodied animals** have many tube feet with tiny suction cups. These feet help them move, feel, and capture food. Spiny-bodied animals are found only in the ocean. If you walk along a beach, you may see some that have washed onto the shore. Some people have collections that include starfish and sand dollars.

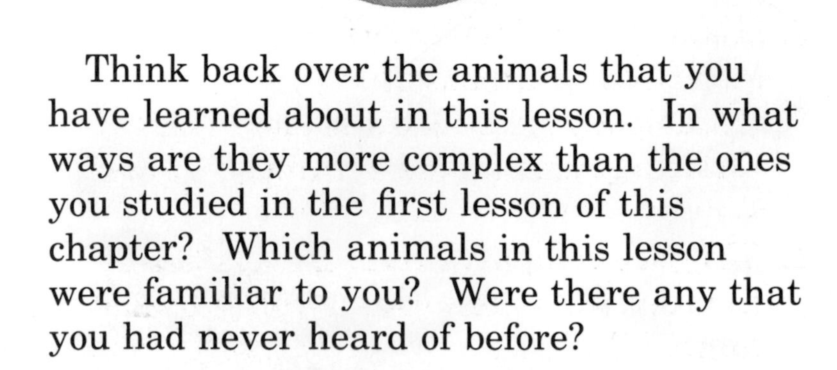

Think back over the animals that you have learned about in this lesson. In what ways are they more complex than the ones you studied in the first lesson of this chapter? Which animals in this lesson were familiar to you? Were there any that you had never heard of before?

A variety of spiny-bodied animals

Lesson Summary

- Complex invertebrates have cells that form tissues and organ systems.
- Mollusks have soft bodies. Many mollusks have outer shells. Segmented worms have bodies that are divided into rings or segments. Arthropods have segmented bodies, legs with joints, and hard exoskeletons. Spiny-bodied animals have skeletons made of plates and tube feet for moving, feeling, and capturing food.

Lesson Review

1. What are the characteristics of mollusks? Name two examples of mollusks.
2. What are the characteristics of spiny-bodied animals?
3. What is an arthropod's skeleton called? What happens as the arthropod grows? Why?

★4. Why are complex invertebrates listed in order from mollusks to spiny-bodied animals?

What are the characteristics of some complex invertebrates?

What you need

1 land snail
1 earthworm
1 cricket
1 hand lens
1 wet cotton swab
covered aquarium
moist soil
pencil and paper

What to do

1. Add the moist soil and animals to the aquarium.
2. Wait a few minutes then observe how each animal moves. Record your observations in a chart.
3. Use the hand lens to observe the body coverings.
4. Gently touch the body of each animal with a moist cotton swab. Record your observations.

What did you learn?

1. How are the body coverings different?
2. Which animals have protection for their soft bodies?
3. How did each animal react to being touched?

Using what you learned

1. Which animal has the most protection against injury to its body organs?
2. How does each animal escape from other animals?
3. Do these animals have well-developed nervous systems? How do you know?

Animal	Type of body covering	How it moves	Reaction to touch
Snail			
Earthworm			
Cricket			

I WANT TO KNOW ABOUT...

Blue Lobsters

Most lobsters found in the Atlantic Ocean are brown or green. They turn red when they are cooked. Sometimes, a blue lobster is trapped. However, these blue lobsters are quite rare. One scientist began breeding blue lobsters so their movements in the ocean could be studied.

The scientist made a surprising discovery. The blue lobsters grew much faster than the brown or green ones. The blue lobsters grew half a kilogram in just under two years. What a discovery! If people could raise blue lobsters on "lobster farms", these lobsters could then be sold. People who like the taste of lobster would have them to eat and the people who raise them could make a good living. Perhaps someday you will see blue lobsters in those tanks at the grocery store.

People who trap lobsters for a living are facing hard times. There is a shortage of lobsters in some areas. This is because too many lobsters have been trapped and sold. This shortage affects the people who depend on the sale of lobsters for their living. There is also a danger that the lobster population will die out completely. Perhaps the blue lobster will be the answer to these problems.

Science and Technology

CHAPTER REVIEW

2

Summary

Lesson 1

- Invertebrates are animals that don't have backbones.
- Sponges are simple invertebrates with many pores through which water flows. Food and oxygen in the water are used by the cells of the sponge. Stinging-celled animals have tentacles that are used for capturing food. Flatworms and roundworms have tissues and organs for moving, digesting food, and getting rid of waste materials.

Lesson 2

- Complex invertebrates have tissues and organ systems.
- Mollusks have soft bodies. Many mollusks have outer shells. Segmented worms have bodies that are divided into rings or segments. Arthropods have segmented bodies, legs with joints, and hard exoskeletons. Spiny-bodied animals have skeletons made of plates and tube feet.

Science Words

Fill in the blank with the correct word or words from the list.

invertebrates	**scavengers**	**regeneration**	**mollusks**
parasite	**spiny-bodied animals**	**sponge**	**arthropod**
flatworms		**stinging-celled animals**	**insects**
roundworms	**segmented worms**		

1. When cells of an animal can regrow into new body parts, this is called ___ .
2. Animals that don't have backbones are called ___ .
3. A(n) ___ is an organism that uses another animal to

supply it with food, water, and oxygen.

4. The largest group of arthropods is ___.
5. A ___ feeds on dead or rotting organisms.

Questions

Recalling Ideas

Correctly complete each of the following sentences.

1. The simplest kind of invertebrate is a
   (a) parasite. (c) sponge.
   (b) mollusk. (d) flatworm.
2. A ___ is an animal with stinging cells.
   (a) jellyfish (c) coral
   (b) hydra (d) all of these
3. There are two openings in the body of a
   (a) roundworm. (c) jellyfish.
   (b) flatworm. (d) planarian.
4. A(n) ___ is a complex invertebrate.
   (a) mollusk (c) segmented worm
   (b) arthropod (d) all of these
5. Complex invertebrates that have bodies divided into rings are
   (a) mollusks.
   (b) arthropods.
   (c) segmented worms.
   (d) spiny-bodied animals.

Understanding Ideas

Answer the following questions using complete sentences.

1. How do flatworms differ from sponges and stinging-celled animals?
2. Name two flatworms.
3. Name three mollusks.
4. What are three examples of segmented worms?
5. What are some characteristics of arthropods?

Thinking Critically

Think about what you have learned in this chapter. Answer the following questions using complete sentences.

1. How are sponges, stinging-celled animals, flatworms, and roundworms similar?
2. Would you expect a sponge or an arthropod to be able to regenerate major body parts more easily? Why?

SCIENCE
In Your World

Vertebrates

This is no ordinary lizard. When in danger, this frilled Australian lizard opens its large mouth and lets out a loud hiss. Then, like an umbrella popping open, it spreads out the colored frill around its head. Although harmless, the frilled lizard looks big and scary enough to frighten away most enemies.

Observed an Open Umbrella?

Look closely at the ribs of an open umbrella. Observe the structure that supports the umbrella. Notice how it is able to open and close.

Now close the umbrella. Then quickly open it again. How do you think the collar on the frilled lizard might be similar to an umbrella?

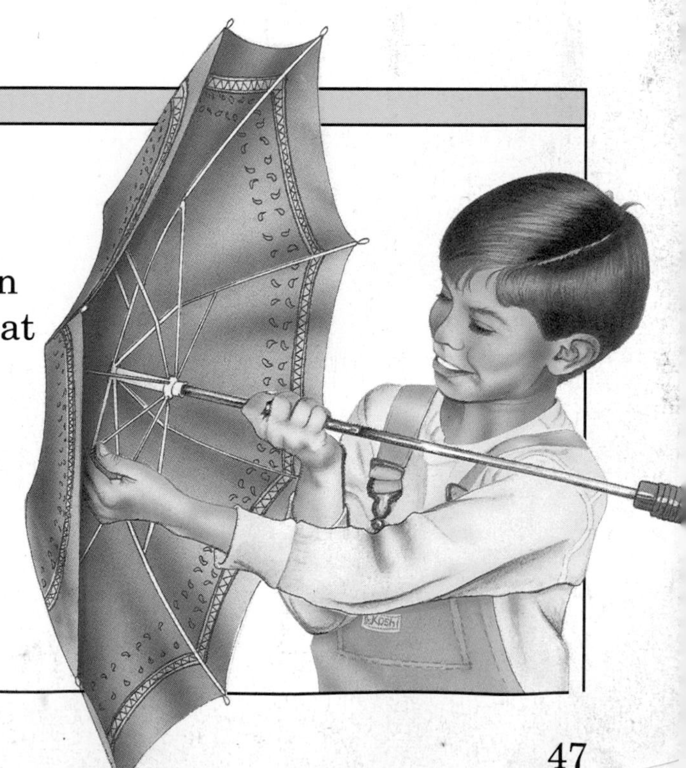

Fish and Amphibians

LESSON 1 GOALS
You will learn
- what vertebrates are.
- what cold blooded means.
- how fish and amphibians are alike and different.

Imagine that you're in a race. As you approach the finish line, something really strange happens. Your backbone disappears. Do you think you'd be able to finish the race? Of course, that couldn't really happen, but just thinking about it gives you an idea of how important your backbone is to the way your body is put together.

What are vertebrates?

In this chapter you will learn about animals that have backbones. Animals with backbones are called **vertebrates.**

Inside its body each vertebrate has a skeleton that shapes the body and protects body parts. An important part of the skeleton is the backbone. Your backbone supports your body and keeps you from being a shapeless mass of muscles.

The skeleton is mostly bone tissue. However, some parts of the skeleton contain cartilage (KART uh lihj). **Cartilage** is a firm, flexible material that covers the ends of most bones.

Skeletons give shape and support to vertebrates.

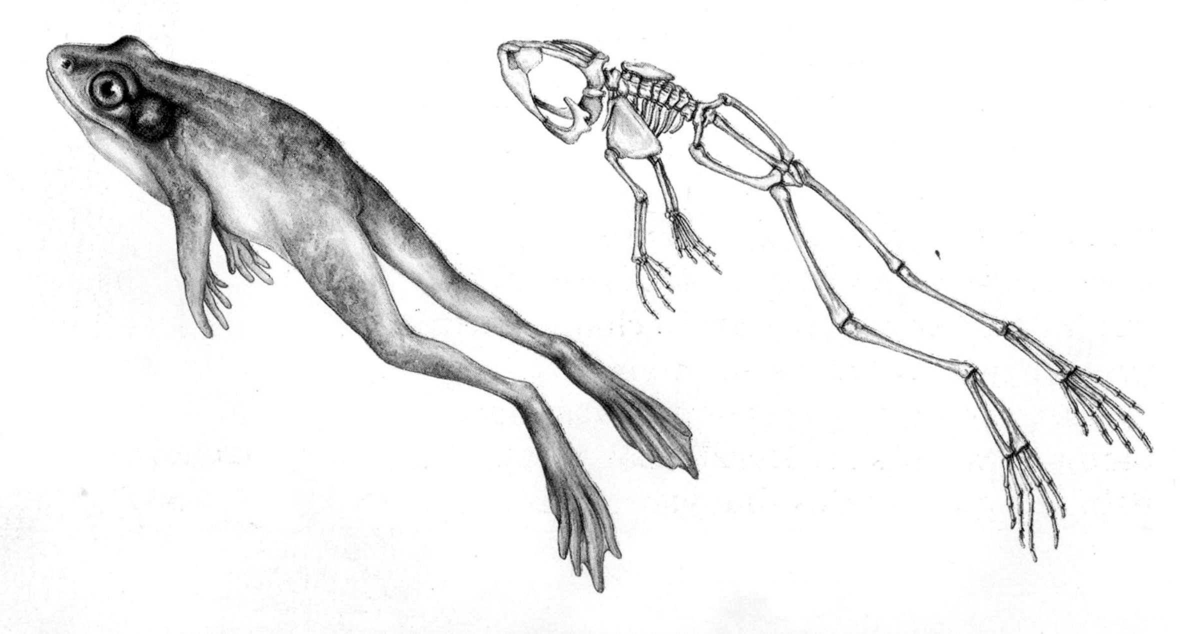

Your skeleton has some cartilage and many bones. Feel your ear. Compare the way it feels to the way your arm feels. Which is more flexible, your ear or your arm? Some fish have skeletons that are made mostly of cartilage.

Scientists classify vertebrates into seven main groups. Each group has special characteristics that are different from those found in the other groups. In this lesson you will learn about fish and amphibians.

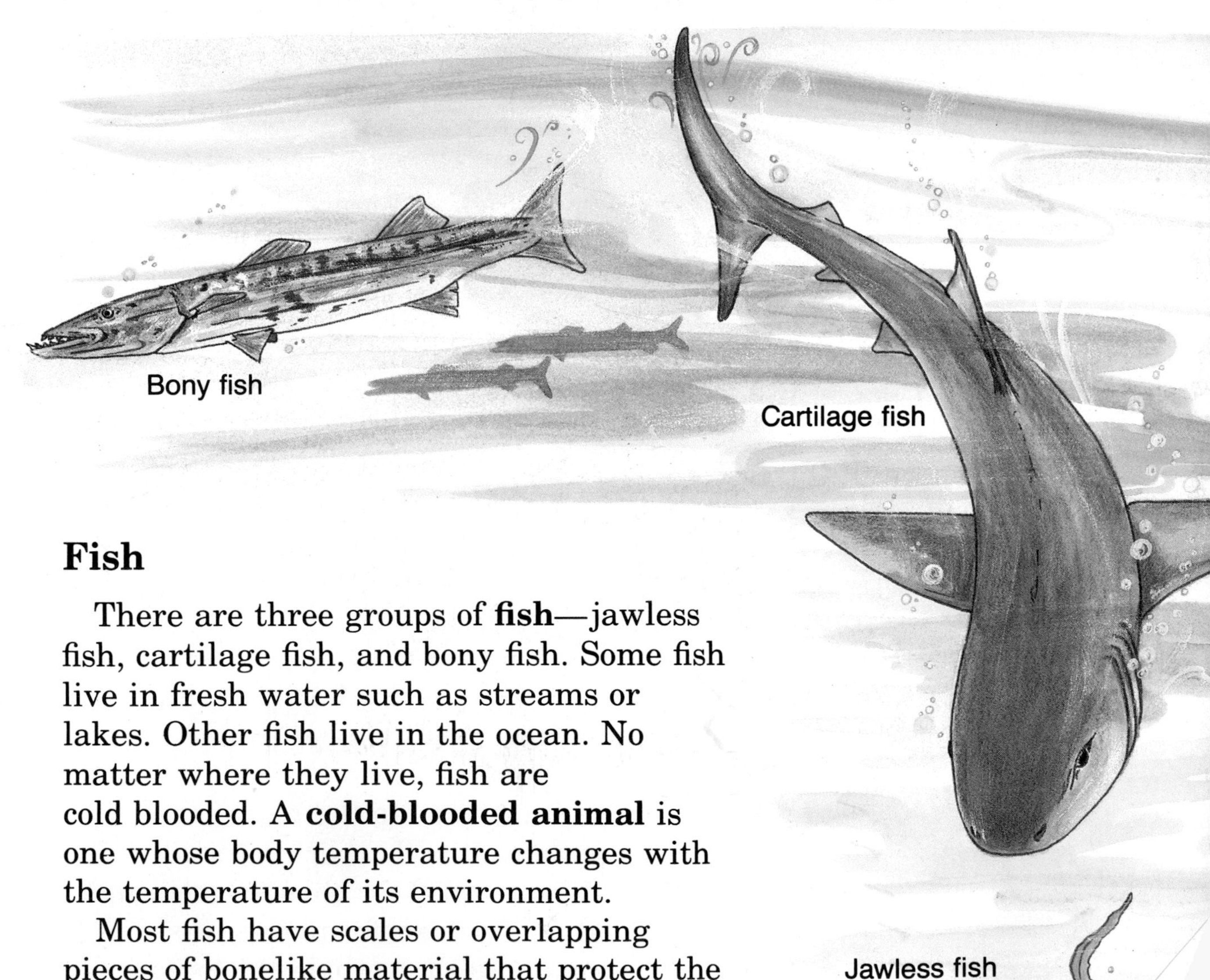

Three types of fish

Fish

There are three groups of **fish**—jawless fish, cartilage fish, and bony fish. Some fish live in fresh water such as streams or lakes. Other fish live in the ocean. No matter where they live, fish are cold blooded. A **cold-blooded animal** is one whose body temperature changes with the temperature of its environment.

Most fish have scales or overlapping pieces of bonelike material that protect the fish. How are scales like your skin?

Most fish have bony structures called fins that help them move through the water the way a rudder and oars help steer a boat.

Gills help fish breathe.

How does a fish get oxygen?

Fish need oxygen to live just as you do. However, fish get oxygen from the water, not from air. A fish gets the oxygen it needs when water flows into its mouth and over and out past the gills. Blood moves through the gills and carries oxygen to the fish's body and carbon dioxide from the body. As water leaves the gills, it also carries away the carbon dioxide.

Most fish reproduce by laying eggs. Although thousands of eggs may be laid at one time, many eggs don't hatch or develop into adult fish. What do you think happens to them?

Fish are important to the environment. They provide food and other materials. Parts of fish are used for chicken feed, in paint, and as fertilizer for plants. Fishing is also a business that provides jobs for many people.

Amphibians

Like fish, **amphibians** are also cold-blooded vertebrates. But they are different from fish in some important ways. Fish live all of their lives in water, but amphibians live part of their lives in water and part on land. They have moist skins and no scales. Most young amphibians live in water, breathe with gills, and have tails. As they get older, they change into adults that breathe with lungs. Frogs, toads, newts, and salamanders are amphibians. Which of these animals have you seen?

What are examples of amphibians?

Amphibians lay many eggs in water or in moist places. Eggs that survive develop into young amphibians with gills and tails.

Young frogs are called tadpoles. Tadpoles change a lot as they grow. Their gills and tails disappear and lungs and legs form. The tadpoles change into adult frogs. The change from young to adult through different-looking body forms is called **metamorphosis** (met uh MOR fuh sus).

Amphibians are an important part of the environment. They eat a lot of flies, mosquitoes, and other insects. Some larger amphibians eat fish, small birds, and other small animals.

Frog metamorphosis

Remember, even though fish and amphibians are different in many ways, they are alike in two important ways. First, they are vertebrates because they have backbones. Second, they are both cold-blooded animals.

Lesson Summary

- Animals with backbones are vertebrates.
- A cold-blooded animal's body temperature changes with its environment.
- Fish are cold-blooded vertebrates that have scaly skins, gills, and live in water.
- Amphibians are cold-blooded vertebrates that live part of their lives in water and part on land.

Lesson Review

1. What characteristic determines whether an animal is classified as a vertebrate or an invertebrate?
2. What is cartilage?
3. What is one way most young amphibians are different from adult amphibians?

★4. If you raised a tadpole, what changes would you see as it grew?

How does temperature change affect a fish?

What you need

goldfish
jar or bowl
water (room temperature, ice water)
thermometer
watch with a second hand
fish net
measuring cup
pencil and paper

What to do

1. Pour 2 cups of room-temperature water into a jar. Record the temperature of the water.
2. Put the fish into the jar. Wait 5 minutes.
3. Observe the fish. Count how many times its gills open in 15 seconds. Multiply this number by 4 to find the number of gill openings in one minute. Record your answer.
4. Remove 1/3 cup of water from the jar and pour in 1/3 cup of ice water. Record the water temperature.
5. Wait again for 5 minutes. Repeat steps 3, 4, and 5.

What did you learn?

1. How did the number of times the gills opened change?
2. What happened to how the fish swam?

Using what you learned

1. What effect does cold temperature have on the activities of cold-blooded vertebrates?
2. How do many cold-blooded vertebrates survive during winter?

Reptiles

LESSON 2 GOALS
You will learn
- the four main groups of reptiles.
- how reptiles are different from fish and amphibians.

If you're like most people, the first thing you think of when you hear the word *reptile* is snake. And you're right—snakes are reptiles. But did you know that there are other groups of animals that are also called reptiles? In this lesson, you'll learn what they are and how reptiles are different from fish and amphibians.

In the first lesson of this chapter, you learned that a vertebrate is an animal with a backbone. You also learned that a cold-blooded animal is one whose body temperature changes with the temperature of its environment. A **reptile** is a cold-blooded vertebrate with dry, scaly skin. Most reptiles have two pairs of legs with five toes on each foot. The toes have claws that help the reptile catch food that it needs to survive. Reptiles run, crawl, climb, or paddle to move around. The four main groups of reptiles living today are turtles, lizards, alligators, and snakes.

These reptiles live in a warm environment.

Reptiles are more complex than amphibians. Their lungs are more developed than the lungs of amphibians. Also, blood takes a more complex pathway through the body in reptiles than in fish or amphibians.

Most reptiles reproduce by laying eggs on land. These eggs have tough, protective shells. They are usually buried in sand, soil, or rotting plant material. As the sun warms the eggs, they hatch into small reptiles that look very much like their parents. Unlike amphibians, reptiles don't go through metamorphosis after birth. Reptiles just grow larger.

Turtles are reptiles with a hard shell that protects them. However, it slows them down. Some turtles live on land, some in fresh water, and some in the ocean. Most turtles are small animals, but some land and sea turtles are huge. Turtles aren't fussy eaters. They feed on plants, worms, insects, and other small animals. Some water turtles also eat fish and birds.

Lizards are another group of reptiles. Some lizards live where it's cold. Others are found in warm, rainy places, and still others live in dry deserts. Lizards eat small insects, animals, or plants. The skin color of many lizards changes to match the color of the area around them. This helps lizards hide from other animals.

Alligators and crocodiles are the largest reptiles. They live in warm, swampy places along streams, rivers, and lakes. Alligators have never seemed very friendly, but they are good parents. They guard their nests carefully so that other animals won't come along and eat their eggs.

Snakes swallow food whole.

Of course, snakes are reptiles too. Some snakes live on land; others live in water. As you know, they don't have legs like other reptiles. Snakes move by catching their scales on the ground and then pushing themselves forward with their muscles. Unlike alligators or crocodiles, snakes don't chew their food; they swallow it whole. Many people think snakes are wet and slimy, but they're not. Their bodies are covered with dry scales.

How do snakes move along the ground?

Dinosaurs are a group of reptiles that died out millions of years ago. There were many different kinds of dinosaurs, both large and small. Can you name any dinosaurs? Do you know why they died out?

In this lesson you have learned about many different animals. Some are big and some are little. Some have legs and some don't. Some have a hard shell and some have soft skin. In spite of all these differences, turtles, lizards, alligators, and snakes are all cold-blooded vertebrates. And they are all reptiles.

Lesson Summary

- Turtles, lizards, alligators, and snakes are the four main groups of reptiles.
- Reptiles have dry, scaly skins, live on land, and are the most complex of the cold-blooded animals.

Lesson Review

1. How do most reptiles reproduce?
2. How are snakes different from other reptiles?

★3. What is one reason reptiles are more complex than amphibians?

Warm-blooded Vertebrates

LESSON 3 GOALS
You will learn
- what warm-blooded vertebrates are.
- how warm-blooded vertebrates are alike and different.

Have you ever had your temperature taken? Your body temperature stays about the same most of the time. It even stays the same when the temperature outdoors changes from very warm to icy cold.

Earlier in this chapter, you learned about cold-blooded vertebrate animals. You know that animals such as fish and snakes have body temperatures that change depending on their environment. Another group of vertebrates are warm-blooded animals. Birds and mammals are **warm-blooded animals** because their body temperature stays the same even if the temperature of the environment changes. Because humans are mammals, you know that whether you're taking a winter walk or playing on a warm, sandy beach, your body temperature basically stays the same. You are a warm-blooded animal.

Why are birds and mammals called warm-blooded animals?

Your body temperature stays about the same.

Birds

Birds are warm-blooded vertebrates that have feather-covered bodies. Just as skin and hair protects your body, feathers protect a bird's body and control its body temperature. A bird's wings and feathers also help it fly and keep its balance.

How are feathers important to birds?

Feathers help birds in flight.

There are many kinds of birds, and they live in very different environments. For example, penguins live in cold, icy areas of the Antarctic. On the other hand, parrots are birds that live in warm, wet places. Some birds, such as geese, ducks, and robins, live in different places at different times of the year. When the weather turns cold in the fall, many of these birds fly, or migrate, to warmer places. How many different birds can you name that live near you? Do any of them migrate?

Many birds build nests to lay their hard-shelled eggs. After the eggs hatch, the adult bird feeds and cares for the young birds until they can fly and feed themselves.

Adult birds are usually very protective of their eggs and their young. Have you ever peered into a nest of eggs or a nest of young birds? Did you ever get squawked at by a parent bird watching you close by?

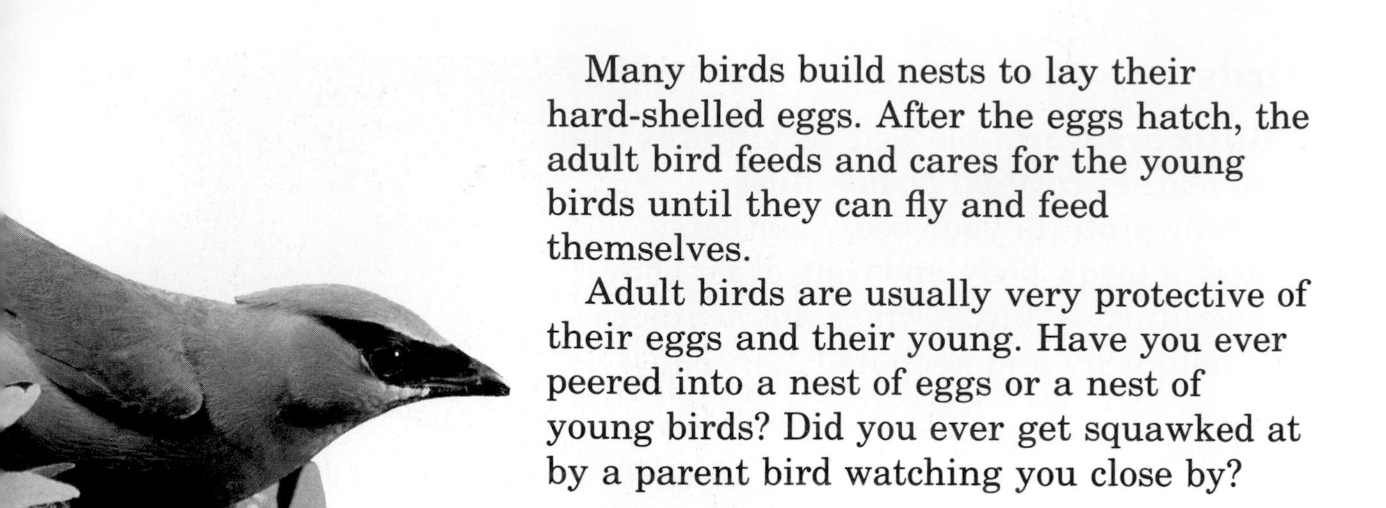

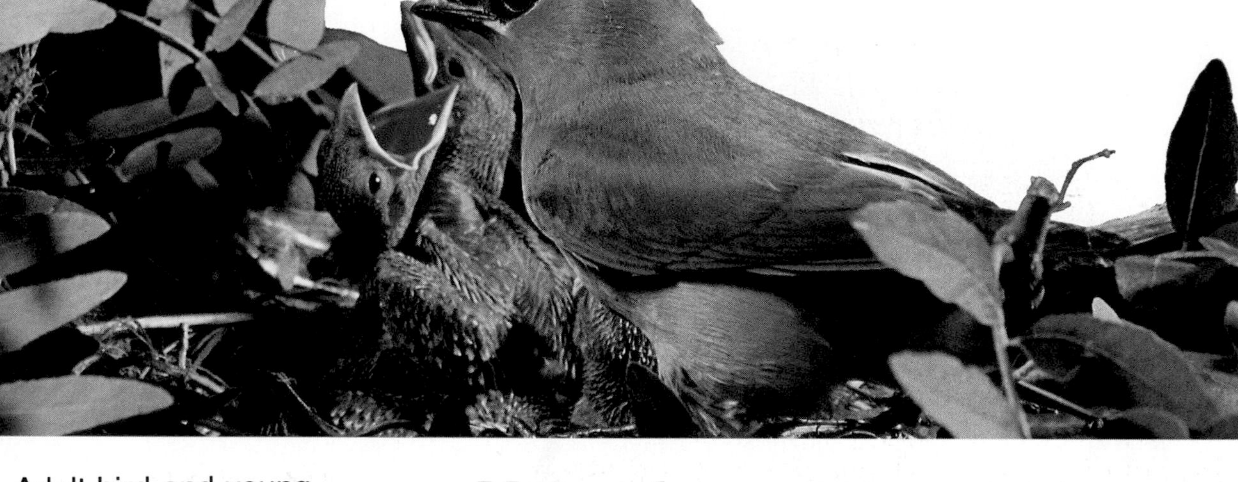

Adult bird and young

Mammals

How are mammals different from birds?

The most complex type of warm-blooded vertebrate is a **mammal.** Mammals include animals such as horses, rats, kangaroos, whales, bats, rabbits, and humans. Unlike birds, mammals are covered with hair or fur. Also, the body of a female mammal produces milk to feed its young. It does not need to hunt for its young.

Like birds, mammals live in different environments all over Earth. What kinds of mammals do you know that live in extremely cold regions? Dry, hot deserts? Warm, tropical areas?

Kangaroos and platypus

Mammals produce their young in different ways. Some mammals, like the platypus and the anteater, lay their eggs in shells. A **marsupial** (mar SEW pee uhl), such as a kangaroo, produces young that develop in a special pouch in the mother. Marsupial eggs develop inside the mother's body, and then they finish growing in their mother's pouch. Other mammals, such as whales, cows, and humans, develop in the mother's body and are born when they are completely formed.

How is a marsupial different from most other mammals?

Mammals have complex systems of organs. Some mammals are able to walk upright and grasp things with their hands and sometimes with their feet, too. Humans are the most complex mammals. You use your brain to think, imagine, learn, question, and understand. You make decisions and solve problems.

Would You Believe?

Pigs and sea lions are two of the few known mammals that sunburn (other than humans).

Certain muscles and joints in your hands allow you to hold objects between your thumb and fingers. These muscles and joints allow you to hold your pencil when you write. To see how important these muscles are, try doing normal activities without using your thumbs. What problems do you have?

In this chapter, you have learned about many different kinds of cold-blooded and warm-blooded vertebrates. Look around you while you are taking a walk or visiting the zoo. Now you will be more aware of the kinds of animals you see and what is different or what is the same about each one.

ACTIVITY

You Can...

Stay Warm With Human "Fur"

Fur protects some mammals from the cold. Humans wear clothes to stay warm. Get four types of cloth. Predict which will make the warmest clothes. Wrap a piece of your first choice around a thermometer and place it in a refrigerator. After 10 minutes record the temperature. Repeat this with your other choices. Which works the best?

SCIENCE AND . . .
Reading

Humans are able to use objects as tools because —

A. they are warm blooded.
B. they care for their young.
C. their hearts have four chambers.
D. their hands have special muscles and joints.

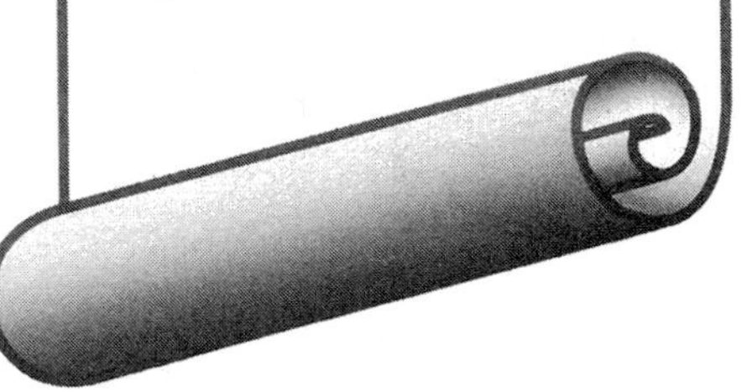

Lesson Summary

- Warm-blooded vertebrates have a body temperature that stays the same even if the temperature of their environment changes.
- Birds are warm-blooded vertebrates with body coverings of feathers; strong, light skeletons; and wings. Mammals are warm-blooded vertebrates with body coverings of fur or hair. Females produce milk for their young.

Lesson Review

1. How are warm-blooded and cold-blooded animals different?
2. How are feathers useful to birds?

★3. How are marsupials and humans alike? Different?

Use Application Activity on pages 355, 356.

How do bones differ?

What you need

bone A | bone B
hand lens | pencil and paper

What to do

1. Have your teacher break the bones apart.
2. Use a hand lens to observe the inside of each bone. Draw and compare the inside of each bone.

Using what you learned

1. Suppose you were blindfolded and held in your hand a chicken bone and a beef bone of the same size. How would you know which bone belonged to an animal that flies?
2. Which bone do you think is more like your own bones? Why?

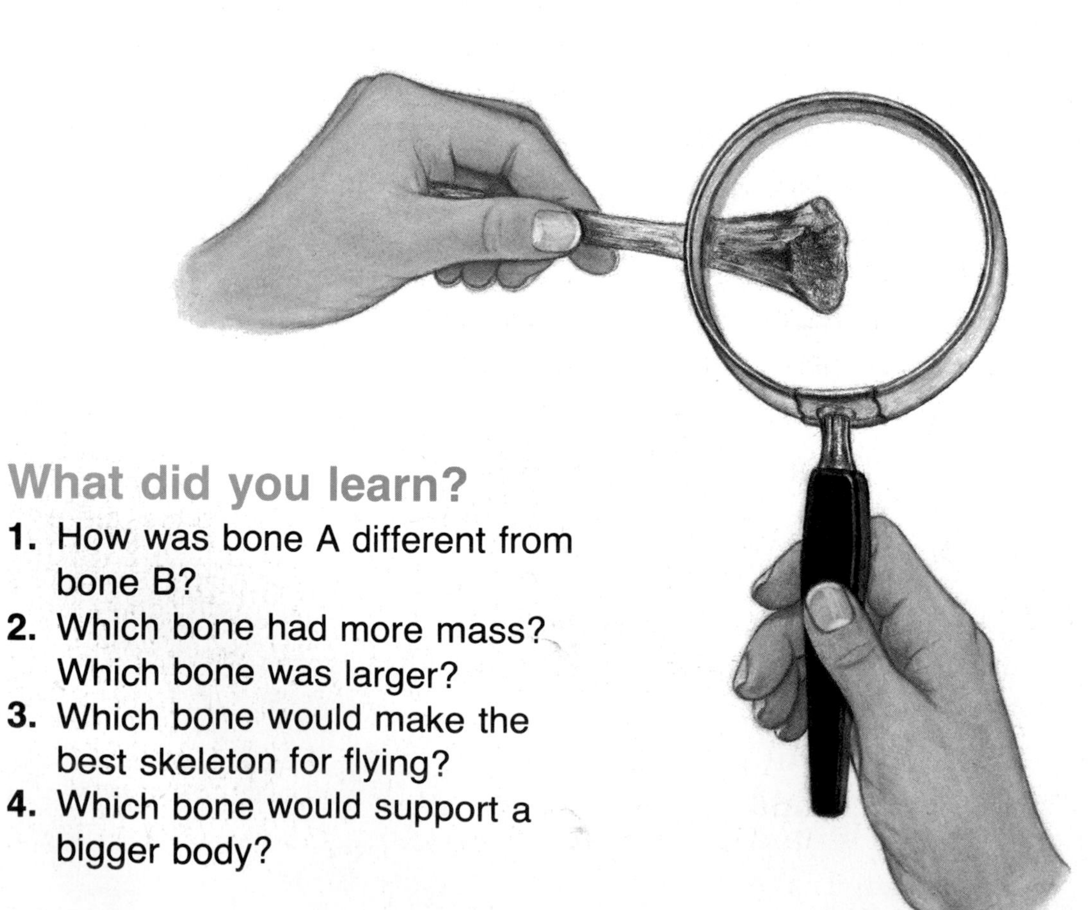

What did you learn?

1. How was bone A different from bone B?
2. Which bone had more mass? Which bone was larger?
3. Which bone would make the best skeleton for flying?
4. Which bone would support a bigger body?

I WANT TO KNOW ABOUT...

An Animal Trainer

Phil DeLeon is training and caring for the dolphins that perform at a large aquarium.

Part of the time, Phil is training the dolphins to learn new tricks. He tosses a fish to a dolphin, and the dolphin lets out a squeak. Phil explains that dolphins squeak when they are upset or excited. He rewards the dolphin for making the sound so that the animal will make it again during the aquarium show. Phil has trained dolphins to do many tricks by using this reward system.

But Phil is also working with the dolphins to do research at the aquarium. Phil is interested in the sounds that dolphins make. Phil thinks they are trying to communicate. What do you think?

As part of the research, Phil records dolphin sounds using underwater microphones. Phil writes down what a dolphin is doing while it is making a certain sound. This research may someday help us to speak to dolphins or to understand what they are saying.

So what do you think? Would you be good at training animals? If you have a lot of love and a lot of patience, the answer could be yes. It's something to think about.

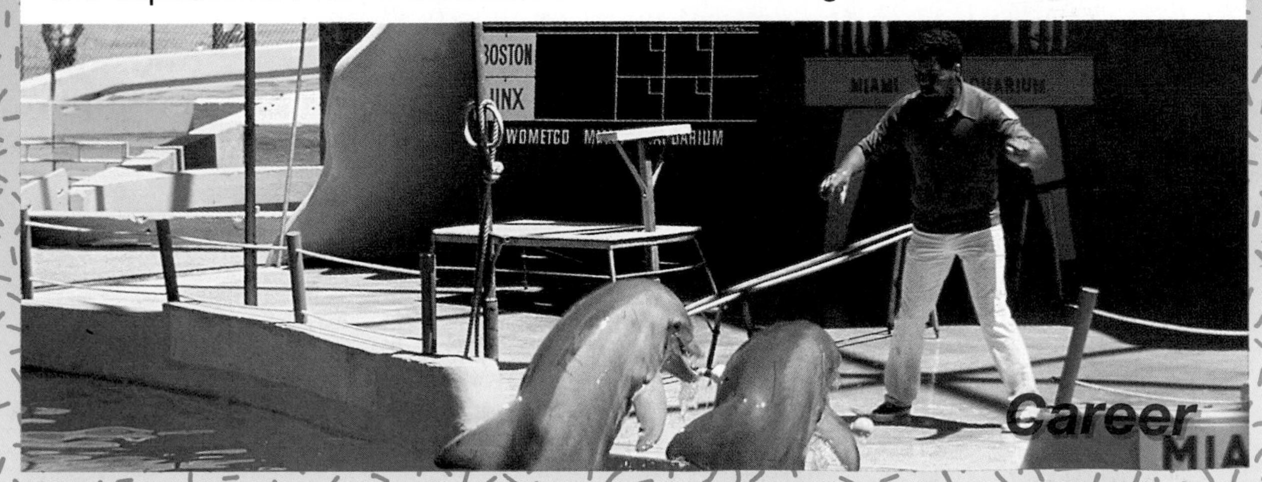

CHAPTER REVIEW 3

Summary

Lesson 1
- Animals with backbones are vertebrates.
- A cold-blooded animal's body temperature changes with its environment.
- Fish have scaly skin, gills, and live in water. Amphibians live in water and on land.

Lesson 2
- Turtles, lizards, alligators, and snakes are four groups of reptiles.
- Reptiles have dry, scaly skin, live on land, and are the most complex cold-blooded animals.

Lesson 3
- Warm-blooded vertebrates have a body temperature that stays the same even if the temperature of their environment changes.
- Birds have feathers; strong, light skeletons; and wings. Mammals have body coverings of hair or fur. Females produce milk.

Science Words

Fill in the blank with the correct word or words from the list.

vertebrates	**metamorphosis**	**amphibians**	**dinosaurs**
cartilage	**warm-blooded animals**	**reptile**	**marsupial**
cold-blooded animal	**fish**	**birds**	
		mammal	

1. A ___ produces young that develop in a special pouch in the mother.
2. The body temperature of a ___ changes with the temperature of the environment.
3. ___ have backbones.
4. Frogs undergo a change from tadpole to adult called ___.
5. Material found at the ends of bones is ___.

Questions

Recalling Ideas

Correctly complete each of the following sentences.

1. A cold-blooded vertebrate with dry, scaly skin is a
   (a) fish. (c) amphibian.
   (b) reptile. (d) all of these.
2. All ___ go through metamorphosis after birth.
   (a) fish (c) reptiles
   (b) amphibians (d) birds
3. The largest reptiles are
   (a) alligators. (c) turtles.
   (b) fish. (d) lizards.
4. Warm-blooded vertebrates with feather-covered bodies are
   (a) mammals. (c)marsupials.
   (b) amphibians. (d) birds.
5. The most complex type of warm-blooded animal is a(n)
   (a) mammal. (c) bird.
   (b) amphibian. (d) reptile.

Understanding Ideas

Answer the following questions using complete sentences.

1. In what ways are amphibians different from fish?
2. In what two ways are fish and amphibians alike?
3. How are reptiles more complex than amphibians?
4. How do snakes differ from other reptiles?
5. What characteristics do birds and mammals have in common?
6. In what ways are birds and mammals different?

Thinking Critically

Think about what you have learned in this chapter. Answer the following questions using complete sentences

1. Why are there fewer snakes in the northern United States than in the southern United States?
2. If you were to find a baby turtle right after birth and make it a pet, what changes would you expect to see as it becomes an adult?

SCIENCE
In Your World

Seed Plants

An Indian tale tells of a tiny man who lives inside a popcorn seed. When you heat his house, he gets so mad that he tears it down. And that is why popcorn pops!

Of course, you know there are no little people living in seeds. But do you also know that in every seed, there is something alive? It's a baby plant, ready and waiting to grow.

Have You Ever...

Examined a Seed?

Carefully split open a lima bean. At one end, you will find a small bump. Look at it with a hand lens. It is the embryo, or baby plant. The rest of the seed is stored food that the baby plant will use when it starts to grow. Now put three beans and a moist towel into a plastic bag for a week. What will happen?

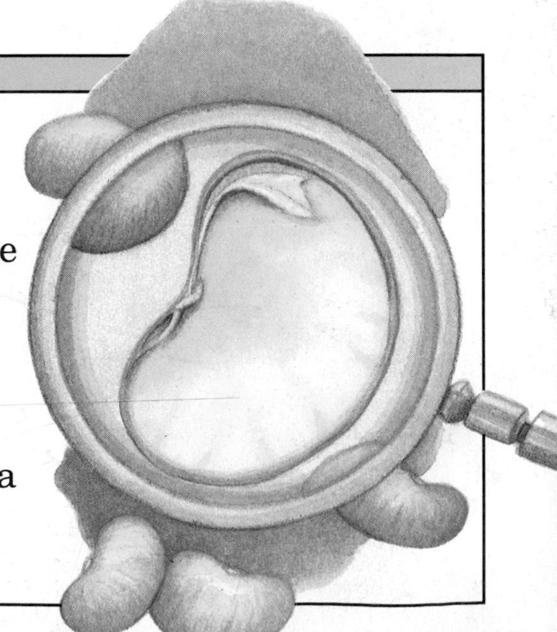

Observing Plants

LESSON 1 GOALS

You will learn

- that plants can be divided into two main groups.
- which major group seed plants are in and how they reproduce.
- that seed plants have different characteristics and grow in different environments.

Mike and Sara have been studying how scientists arrange living things into groups. One day they decided to make a list of all the living things they saw on their way home from school. Their list included these things—long-haired cat, red roses, grass, collie dog, ants, evergreen tree, tomato plants, spider, moss, earthworms. How would you group the things Mike and Sara had on their list?

In Chapters 2 and 3, you learned about a large group of living things called animals. You learned that all animals have some similar characteristics. But because some animals have backbones and others don't, we can form two main groups of animals. In the next two chapters, you will learn about another large group of living things—plants. Plants have some characteristics that are alike. They also have a major characteristic that differs, allowing us to group or classify them.

Mike and Sara study living things.

Classifying Plants

From their observations, Mike and Sara found many kinds of plants—roses, grass, an evergreen tree, tomato plants, and moss. If they study plants in more detail, they will learn that plants can be classified into two main groups—those with a system of tubes and those without a system of tubes.

Plants with tubes, such as ferns, trees, and flowers, have systems similar to the blood vessels in your body. Blood vessels carry blood throughout your body. Tubes in these plants carry **nutrients,** such as water and minerals, from the soil to every part of the plant. Think of the tremendous height of some trees. The height, or distance between roots and leaves, can occur because of the carrying ability of the tubes.

Plants with and without tubes

Some plants don't have a system of tubes. Seaweeds and mosses belong to this group. These plants without tubes need to live in or near water or where it is very moist. These plants do not grow to the heights that we notice in plants that have a system of tubes.

Think of the kinds of plants you see most often. You might think of trees from which lumber is made or carrots and other vegetables you eat during meals. These plants are similar in two ways. First, they have a system of tubes. And second, these plants can reproduce with plant parts called seeds. Some of these plants form their seeds inside flowers. Other plants, such as pine trees, form their seeds inside cones. All plants that have a system of tubes and form seeds are **seed plants.**

Comparing Seed Plants

From our examples of trees and carrots, you can see that seed plants have different characteristics. Seed plants grow to be different sizes and have different colors and shapes. These different characteristics make it possible to tell one plant from another. How can you tell a maple tree from an elm tree, or a sunflower from a rose?

Seed plants grow in different environments. For example, cattails grow in wet places while other seed plants, such as cacti, live well in dry areas. Some seed plants grow in sunny, warm places. Others grow where there is shade and cool temperatures. Each type of plant has characteristics that enable it to grow in a particular environment. For example, the cacti in the picture store water, so they can live where there is little rain. What kind of seed plants grow well in your neighborhood? How are they different from other seed plants?

SCIENCE AND . . .
Math

Carlos bought 48 young plants for his greenhouse. Which of the following patterns might Carlos use for planting?

A. 6 rows of 8 plants
B. 6 rows of 7 plants
C. 4 rows of 9 plants
D. 8 rows of 7 plants

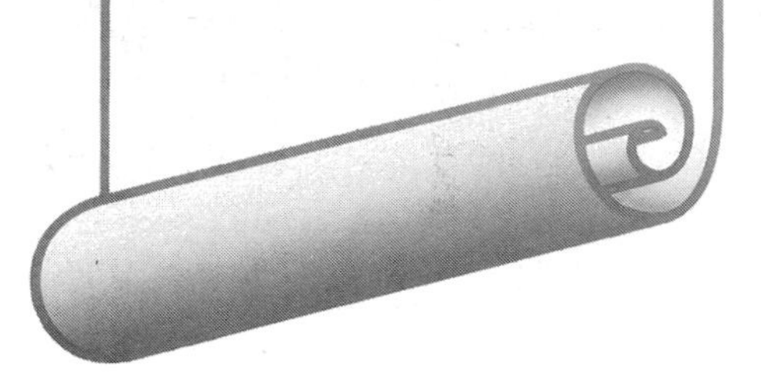

Seed plants in different environments

ACTIVITY

You Can...

Make a Seed Study

You eat many kinds of seeds without thinking about it. Survey foods in your home. List those that come from seeds. Make a poster showing ten of the foods and their seeds. Find these countries on a world map: Zimbabwe, Mexico, Canada, China, and the Soviet Union. Which seeds are important food sources in these countries?

Lesson Summary

- There are two large groups of plants—those with a system of tubes and those without a system of tubes.
- Seed plants have a system of tubes and reproduce by forming seeds.
- Seed plants grow in different colors, shapes, and sizes and grow in wet, dry, warm, and cool environments.

Lesson Review

1. What does "classifying plants" mean?
2. What parts of seed plants can form seeds?

★3. Why does a plant live in a particular environment?

Use Application Activity on pages 357, 358.

What are some different kinds of seed plants?

What you need

flower (cut in half)
pine cone
hand lens
1/2 peach with pit
1/2 apple
newspaper
pencil and paper

What to do

1. Your teacher will cut the flower in half. Use a hand lens. Study the part of the flower where tiny seeds develop.
2. Now study a pine cone. Find where seeds are produced. Shake the pine cone and observe the thin, flat seeds.
3. Look at the peach half and apple half. Count the number of seeds you see.

What did you learn?

1. In what parts of seed plants are seeds formed?
2. Name a fruit that has only one seed. Name another fruit that has several seeds.

Using what you learned

1. There are many different kinds of flowers. How are all flowers alike?
2. Name two other fruits that have one seed. Name two that have many seeds.

Structure of Seed Plants

LESSON 2 GOALS
You will learn
- how plant cells are different from animal cells.
- why roots, stems, and leaves are important parts of plants.

You learned in Chapter 2 that animals are made of small units called cells. Animal cells are too small to be seen except with a microscope. These cells, however, carry out the activities that keep the animal alive. Plant cells are very much like animal cells, except that plant cells have cell walls that give them their shape. To get an idea of what cell walls are like, imagine a sturdy shoe box with a lid. Inside is a plastic bag filled with water. The bag is tied shut. When you put the bag into the box, the bag is shaped and held in place by the sides of the box. The sides of the box are like the cell walls of plants.

Cells of living things

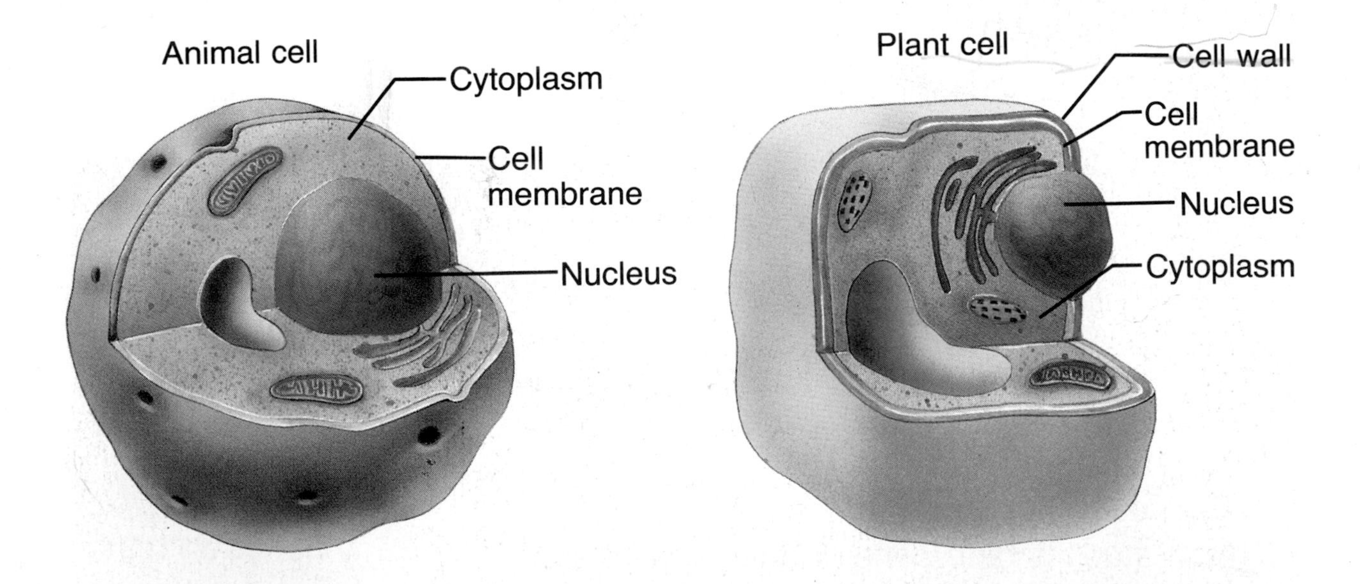

How are all seed plants similar?

You have already learned that seed plants have characteristics that differ. However, most seed plants have the same kinds of parts. These parts are roots, stems, and leaves.

Roots

Roots are the parts of a plant that bring nutrients into the plant. Tiny root hairs grow on the outside of each root. These root hairs absorb the nutrients that a plant takes in from the soil. Once inside a root, nutrients move up the tubes to the rest of the plant.

Roots also hold plants in the ground. A root's ability to support a plant depends on its size and depth in the ground. Some plants, like carrots, have long, thick roots that have the plant's food stored in them. Other plants, like grasses, have small roots that look like branches or threads.

Would You Believe?

Plants may be grown without soil by a process called hydroponics in which the exposed roots are soaked in water containing nutrients.

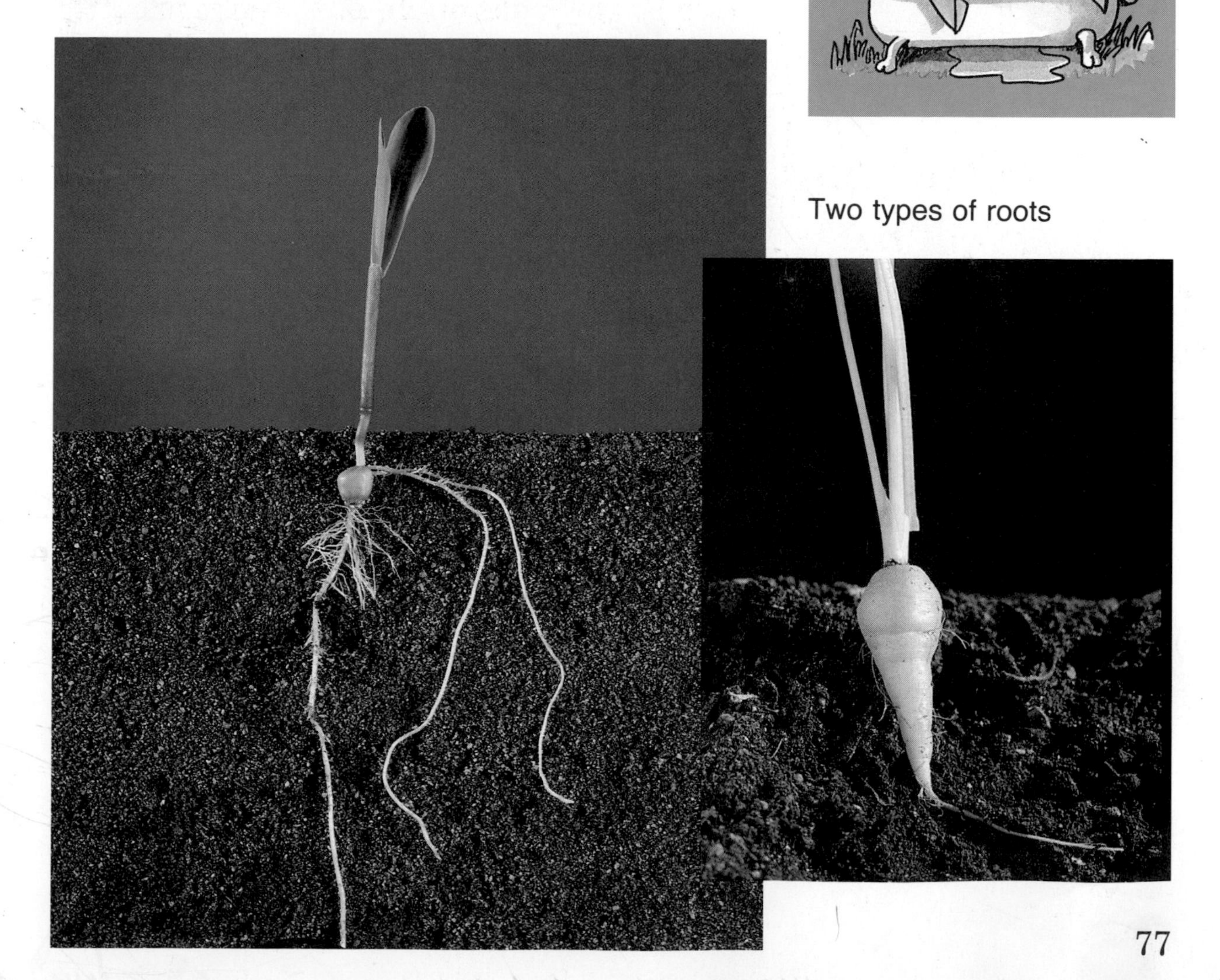

Two types of roots

Stems

Stems are the parts of a plant that contain tubes that transport or carry materials between the roots and leaves. The nutrients move up from the roots and into the leaves of the plant. Stems also support the plant's leaves. Tomatoes, beans, and tulip plants have soft, green stems. Maple trees and rosebushes have hard, woody stems. Not all stems look skinny like the ones we've described. It may surprise you to know that a potato is a kind of underground stem that stores food.

Stems are noticeably different.

Leaves

Most seed plants have **leaves** in which food is made. Leaves have small tubes called veins. Water and other nutrients move through veins just like they move through the tubes in roots and stems. The water is used by leaves in the process of making food.

Leaves come in different sizes and shapes. Some leaves, such as those of a pine tree, are narrow and look like needles. Other leaves are broad and flat like maple leaves. Leaves like those of the maple tree with only one part are called simple leaves. Broad leaves with leaflets, such as hickory leaves, are compound leaves. Which leaves have you seen most often?

Simple and compound leaves

Lesson Summary

- Cell walls give plants their shape.
- Roots take in nutrients from the soil and hold the plants in the ground. Stems hold up the leaves and carry nutrients from roots into the rest of the plant. Leaves are the parts of the plant in which most food is made.

Lesson Review

1. Compare the two types of roots found in seed plants.
2. What carries nutrients and water through stems of a seed plant?

★3. Compare the two types of broad leaves.

How does water move to leaves?

What you need

jar
water
blue food coloring
spoon
fresh celery stalk with leaves
crayons
pencil and paper

What to do

1. Fill the jar half full of water. Add six drops of food coloring and stir.
2. Observe the freshly cut end of the celery. Draw and color how the bottom of the stalk looks. Place it into the jar of colored water.
3. Wait two hours and observe the celery. Record your observations.
4. Take the celery out of the water. Your teacher will cut across the bottom and top of the stalk.
5. Observe the cut ends of the celery. Draw and color how the bottom of the celery looks now.

What did you learn?

1. How did the ends of the celery look in steps 2 and 5?
2. Describe what happened to the leaves.
3. What happened in the tubes of the celery stalk?

Using what you learned

1. How do you know that the water went up the celery stalk?
2. How does water get to the celery stalk when it is growing in the garden?
3. Why do leaves and flowers wilt when the stem is cut?

I WANT TO KNOW ABOUT...

Working in a Greenhouse

Have you ever seen a plant with yellow, orange, red, and green leaves? There is such a plant. It is called a croton (KROH tahn). There is even a plant with striped leaves called a zebra plant. Plants like these can be found in the greenhouse at Gina's Garden Center.

Gina does various things in the greenhouse. She takes care of young trees, flowers, and plants. She makes sure the young plants get the right amount of light, water, and fertilizer. She also makes sure that the greenhouse stays at the right temperature.

Before selling a plant to a customer, Gina very carefully explains how to take care of the plant. She talks to her customers about when to water the plants and how much water to give them. She also tells how much sunlight the plant needs, and whether or not it is poisonous. Gina wants to be sure each customer knows what he or she is buying and that her plants are going to have a good home.

Career

CHAPTER REVIEW 4

Summary

Lesson 1

- There are two groups of plants—those with a system of tubes and those without a system of tubes.
- Seed plants have a system of tubes and reproduce by forming seeds.
- Seed plants are different colors, shapes, and sizes and grow in many environments.

Lesson 2

- Cell walls give plants their shape.
- Roots take nutrients from the soil and hold the plants to the ground. Stems hold up the leaves and carry nutrients from the roots into the rest of the plant. Leaves are the parts of the plant in which most food is made.

Science Words

Fill in the blank with the correct word or words from the list.

nutrients **stems** **seed plants**
roots **leaves**

1. ___ bring nutrients into the plant.

2. Plants make food in ___.

3. ___ transport or carry materials between the roots and leaves.

4. Water and materials from the soil are called ___.

Questions

Recalling Ideas

Correctly complete each of the following sentences.

1. ___ are plants with tubes that carry nutrients from the soil throughout the plant.
   (a) Trees (b) Mosses (c) Seaweeds (d) All of these
2. ___ are plants that don't have tubes to carry nutrients.
   (a) Ferns (b) Mosses (c) Trees (d) Flowers
3. All plants that have a system of tubes and form seeds are called
   (a) seaweeds. (b) algae. (c) seed plants. (d) mosses.
4. Seed plants reproduce with parts called
   (a) leaves. (b) seeds. (c) stems. (d) roots.
5. ___ form seeds inside cones.
   (a) Pine trees
   (b) Flowering plants
   (c) Seaweeds
   (d) Mosses
6. Plants are held in the ground by
   (a) leaves. (b) seeds. (c) roots. (d) stems.
7. ___ support leaves.
   (a) Roots (b) Stems (c) Seeds (d) Flowers
8. Leaves have small tubes called ___ through which nutrients move.
   (a) cell walls (b) stems (c) flowers (d) veins

Understanding Ideas

Answer the following questions using complete sentences.

1. What are two main groups of plants?
2. What are some different characteristics of seed plants?
3. How are vegetables and trees similar?
4. What three parts do most seed plants have?
5. How are plant cells different from animal cells?
6. What determines a root's ability to support a plant?

Thinking Critically

Think about what you have learned in this chapter. Answer the following questions using complete sentences.

1. You are planning a garden. What do you need to know about the seed plants you plan to grow?
2. Would you expect to see mosses growing in the desert? Explain your answer.

SCIENCE
In Your World

Plants Grow and Change

A walk through weeds can be fun. But have you ever come out of the weeds with seeds sticking to your clothes or to the fur of your pet? One careful observer noticed how some seeds attach themselves. Guess where he got the idea to create the fastener that you know as Velcro?

Have You Ever...

Wondered How Seeds Travel?

Use a hand lens to look at the way Velcro is made. Feel each piece. One piece is fuzzy. Pinch the other strip and look down into it. How do you think Velcro holds together? Now use the hand lens to look at some seeds that stick to clothes. In what ways are these seeds similar to Velcro? How do you think these seeds travel?

Seeds

LESSON 1 GOALS
You will learn
- that seeds have different characteristics.
- that seeds have three important parts.
- that seeds need the right conditions to germinate, or sprout.

Think about times when you have eaten sweet, juicy fruits. Perhaps you had a peach with your lunch, and your meal was interrupted by a large, bumpy pit. Or maybe last summer your thirst-quenching meal of watermelon was interrupted by black or white seeds mixed in with the sweet red pulp. It may surprise you that peach pits and watermelon seeds have something in common. Each of these objects contains a seed that may grow and produce more peaches or watermelons if conditions are right.

Seeds of different sizes, shapes, and colors

You have probably seen other very different kinds of seeds. Some may have been very hard to recognize as seeds. Look at the seeds on this page. What do you notice about each of the seeds? What kinds of plants would you expect to see if each kind of seed became a new plant? Does the size, shape, and color of the seed affect the kind of plant it produces?

Coconut and maple seeds

Seeds are different in many ways. They can be as fine as dust, like orchid seeds, or they can be very large, like coconut seeds. Seeds may be brightly colored, speckled, brown or black. Seeds may also have different shapes. Seeds of burdock plants have small hooks, while those of the maple tree have parts that look like wings.

Some seeds have two halves. Seeds of beans, carrots, peas, and potatoes have two parts. Other seeds, such as corn, dates, and tulips, have only one seed part.

What three parts are found within every seed?

Seeds are also alike in some ways. They are like neat packages with everything a new plant needs wrapped up inside. Each seed contains a tiny plant, stored food, and an outer skin. The young growing plant inside the seed is the **embryo.** The embryo uses the stored food inside the seed when it begins to grow. The outer skin of the seed is the seed coat. It protects the other parts of the seed.

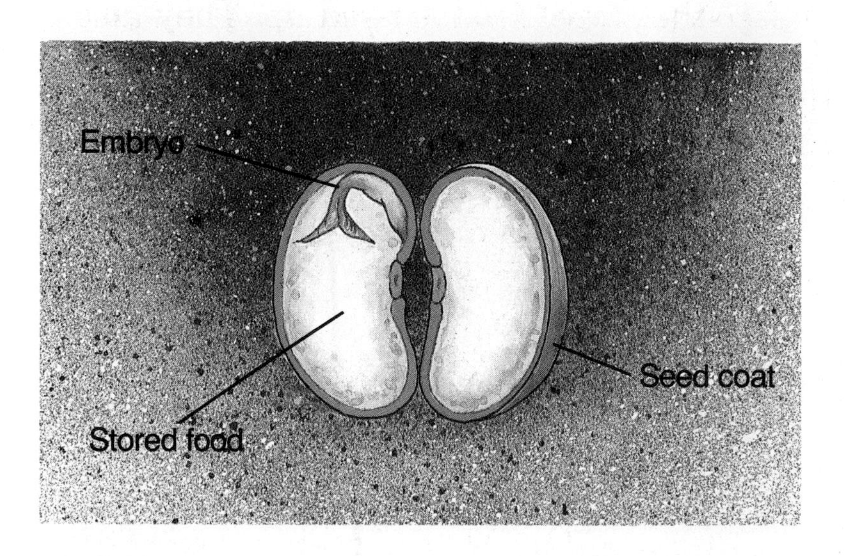

Three parts of a seed

The Growth of Seeds

Most seeds don't grow into new plants right away. Some seeds may stay on the ground for a long time. Other seeds may be carried by the wind or animals to other places. Little plants then begin to grow in places like cracks in the sidewalk. Of course, you can plant seeds in a flower or a vegetable garden. When the seeds have everything they need, they begin to grow, or sprout, into plants.

Germination (jur muh NAY shun) begins when a plant embryo starts to grow. Many seeds germinate in the spring when the temperatures of the ground and the air become warmer. Water from the ground moves through the seed coat and softens it. The seed coat splits as the embryo starts to grow. The embryo grows into a young plant called a seedling.

Seedlings need the right temperature and the right amount of water to grow. Seedlings need air to make their own food after the stored food is used up. They also need the right amount of sunlight for making their own food.

Would You Believe?

Imagine sleeping longer than Rip Van Winkle. Some lotus seeds have germinated after resting for 700 years.

Planting seeds in spring

Fruits

Some seeds grow inside a plant part called a **fruit.** Most of the fruits you see each day are fleshy like a peach or watermelon. Some fruits, however, are very dry when they are ripe. Maple trees and milkweed plants have dry fruits with seeds that are scattered by the wind.

Common fruits

Lesson Summary

- Seeds have different sizes, shapes, and colors.
- A seed has stored food, an embryo, and a seed coat.
- Seeds need water and the right temperature to germinate.

Lesson Review

1. Name four ways in which seeds may be different.
2. How is the seed coat important for a seed?

★3. Why do many seeds germinate in spring?

What are the conditions for seed germination?

What you need

4 small milk cartons
labels
potting soil
bean seeds
water
metric ruler
pencil and paper

What to do

1. Label the milk cartons **Light, Dark, Wet,** and **Dry.**
2. Fill each carton with soil. Plant 4 seeds in each carton. Space the seeds evenly.
3. Put the cartons in these places.
   Light—near a window
   Dark—in a closet
   Wet—on a table
   Dry—on a table
4. Keep the soil moist in each carton except the one labeled Dry.
5. Keep all cartons at the same temperature.
6. After one week, observe, measure, and record the heights of the plants.
7. Wait a second week.

What did you learn?

1. In which cartons did the plants grow? In which cartons did the plants not grow well?
2. What were the results after the second week?

Using what you learned

1. How does light affect the germination of seeds?
2. How does moisture affect the germination of seeds?
3. What is the effect of cold temperature on seed germination?

Plants Make Food

LESSON 2 GOALS
You will learn
- that plants can make food.
- that plants use three important materials to make food.

You may have seen a bumper sticker that said, *Have you thanked a green plant today?* At first you may think that this is a silly question. However, if you know how important green plants really are, the question won't seem so silly.

In what important way are plants different from animals?

Do plants and animals have anything in common? You know that they both need water and nutrients to survive. How, then, are plants different from animals? They are different in that plants can make their own food. Animals depend on plants for food or on other animals that eat plants.

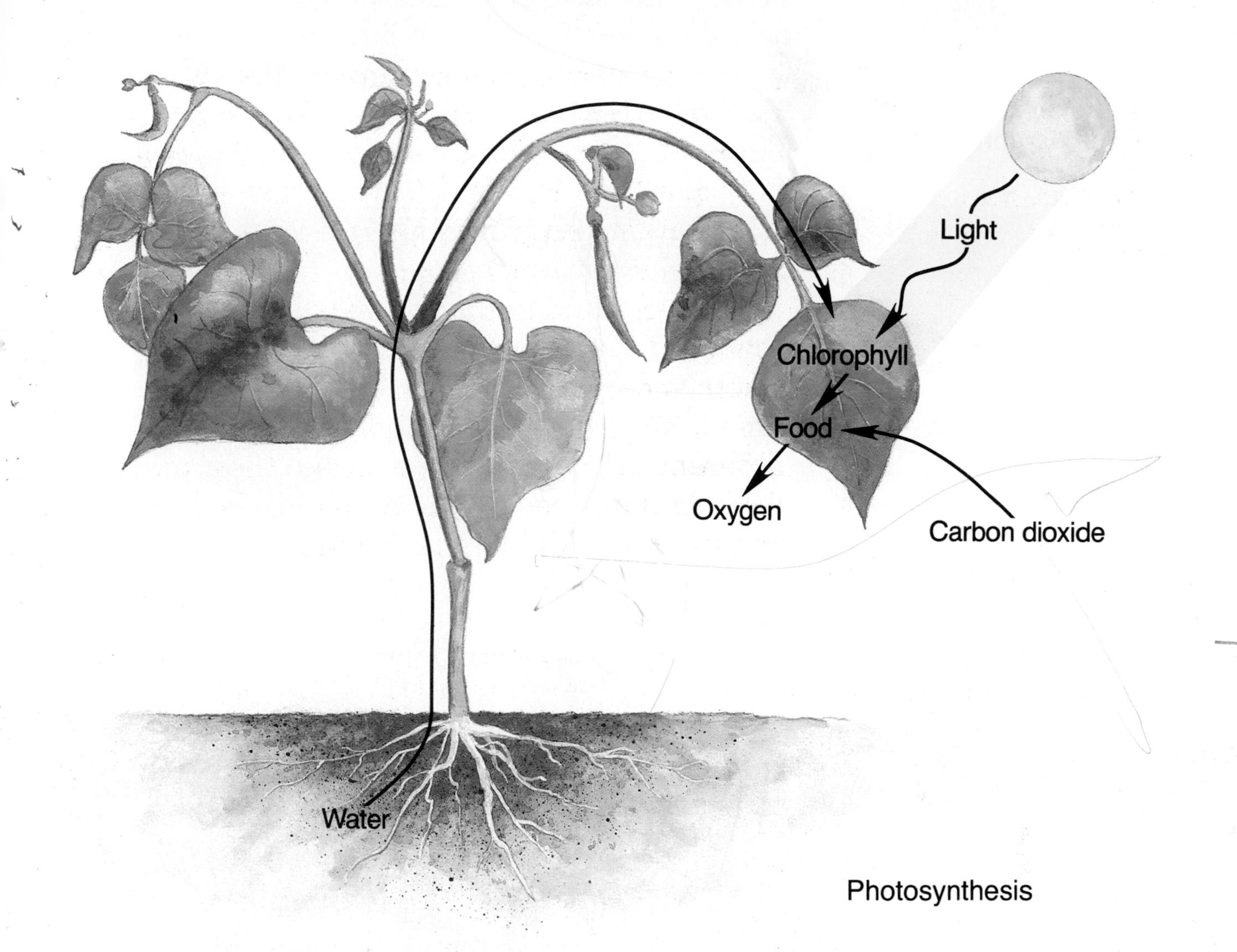

Photosynthesis

The process by which plants make food is **photosynthesis** (foht uh SIHN thuh sus). During photosynthesis, green plants make food in their leaves. This process takes place during the day when there is sunlight. The plants take water from the soil and carbon dioxide from the air. Their leaves have a chemical called **chlorophyll** (KLOR uh fihl) that traps energy from sunlight. In the leaves, this energy is used to change carbon dioxide and water to food.

What is photosynthesis?

Photosynthesis also makes oxygen. So, not only do plants make food, but they provide oxygen for us and other animals. HAVE YOU THANKED A GREEN PLANT TODAY?

Trees are an important source of oxygen.

Lesson Summary

- Photosynthesis is the process by which plants make food.
- Plants use chlorophyll, carbon dioxide, and water to make food.

Lesson Review

1. How is chlorophyll important to photosynthesis?
2. Why doesn't photosynthesis take place at night?
3. ★ What two products of photosynthesis are important to animals?

How can you show that photosynthesis produces starch?

What you need

scissors
geranium plant
black paper
paper clips
shallow dish
alcohol
water dropper
tincture of iodine
tweezers
pencil and paper

What to do

1. Clip a strip of black paper across the upper surface of a geranium leaf. Put the plant in a sunny window for 2 days.
2. Cut the covered leaf from the plant and remove the paper.
3. Soak the leaf in alcohol for 1 day. **CAUTION:** *Do not taste the alcohol.*
4. Use tweezers. Remove the leaf from the alcohol and rinse it off. Drop iodine on both the covered and uncovered parts of the leaf. **CAUTION:** *Do not get iodine on your skin or clothing. Do not taste it.*

What did you learn?

1. Why did you put black paper on the leaf?
2. What happened when the leaf was soaked in alcohol?
3. What happened when you dropped iodine on both parts of the leaf?

Using what you learned

1. Which part of the leaf contained starch?
2. Where does the energy come from that plants use to make food?

IODINE TEST	
Uncovered part of leaf	Covered part of leaf

Making New Plants

LESSON 3 GOALS
You will learn
- that flowers contain important parts.
- that plants can produce new plants in different ways.
- that all seed plants have a plant life cycle.

Juanita helps Mr. Walker at his flower shop, where there are many different kinds of plants. Juanita wants to grow as many kinds of plants as possible. However, to do this she needs to learn more about ways that plants can produce new plants. What are some ways you know for growing new plants?

Juanita knows that flowers are very important parts of plants because they produce seeds. She asks Mr. Walker if he will tell her more about flowers and their parts. Let's listen in as they look at the parts of a flower.

Juanita and Mr. Walker look at flower parts.

The outer parts of the flower that protect the bud of the flower before it opens are the sepals. The parts of the flower that are usually brightly colored are the petals. They surround the inside parts of flowers.

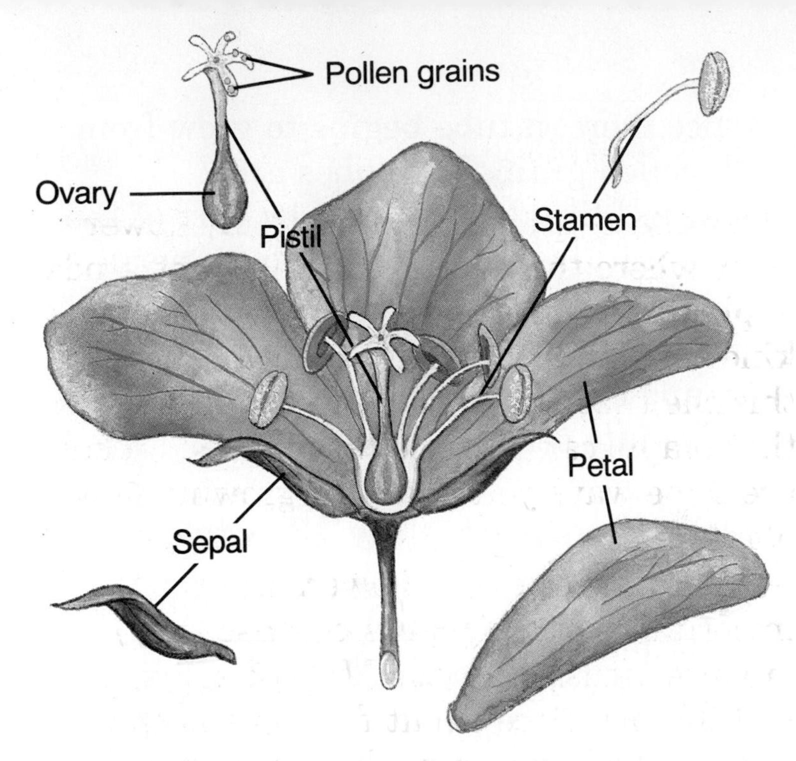

Parts of a flower

Inside the petals of most flowers is the female part of the flower, called the **pistil.** The top of the pistil is narrow and the tip is sticky. The ovary is at the bottom of the pistil. Inside the ovary are egg cells that can become seeds.

Most flowers have **stamens** (STAY munz), which are the male parts of the flower. The stamens produce a powdery material called **pollen.** The pollen particles, called grains, contain sperm cells. New seeds are produced when sperm cells combine with egg cells inside the ovary.

Pollination

In what two ways might pollination occur?

Pollination (pahl uh NAY shun) takes place when pollen grains move onto the sticky part of the pistil. Pollen grains can be moved by wind or carried by insects, such as bees, to the pistil. The pollen grains brush off onto the pistil of each flower as the bee visits flowers.

SCIENCE AND . . .
Writing

As Jill and I walked through the field of wildflowers, we noticed that the honeybees ____ from flower to flower gathering nectar.
A. is moving
B. are moving
C. were moving
D. does move

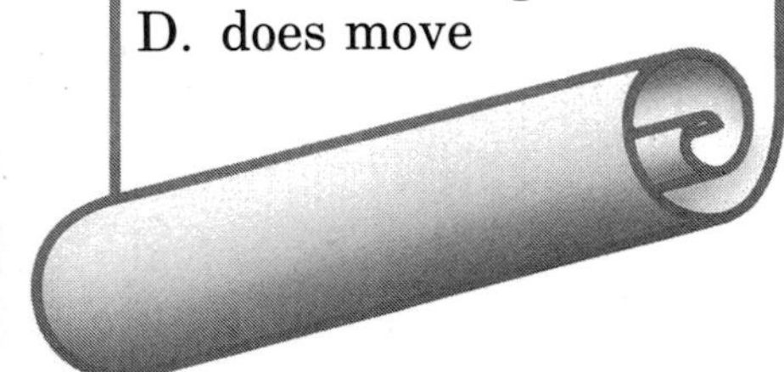

Once there, a tube begins to grow from each pollen grain. The tubes grow downward through the narrow part of the pistil until they reach the ovary. Sperm cells are released from the tube into the ovary. Fertilization occurs when the sperm cells and egg cells join. The fertilized eggs develop into seeds.

Plants Produce New Plants

Juanita had noticed that sometimes Mr. Walker placed the roots of plants in water to begin new plants. Mr. Walker explained that there are other ways to grow plants besides germinating seeds. You can cut a leaf or stem from a plant to start a new plant. When you put a cutting of the plant in water, the plant grows new roots.

Some plants, like daffodils and tulips, have bulbs. The bulbs have food stored inside fleshy leaves. This food is used as the bulb produces a new plant. Other plants, such as grasses, reproduce plants with underground stems called rhizomes (RI zohmz).

New plants grow from cuttings and bulbs.

Some plants, such as white potatoes, have special rhizomes called tubers. These tubers have stored food and can reproduce from "eyes" or buds.

Perhaps you have seen strawberry plants growing along the ground. They have thin stems that form new plants at their tips. These stems are called runners. A new plant grows where the tip of the runner touches the ground.

Plant Life Cycle

After plants form seeds inside fruits or cones, the fruits ripen or the cones open. The seeds are scattered and fall on the ground. If conditions are right, the seeds germinate. The germination, and the growth of a plant, and the formation of new seeds are parts of the **plant life cycle.**

What is the plant life cycle?

The life cycles of some seed plants are completed in one year. The life cycles of other kinds of plants take more than one year. No matter how many years are needed, all seed plants go through a plant life cycle.

Plant life cycle

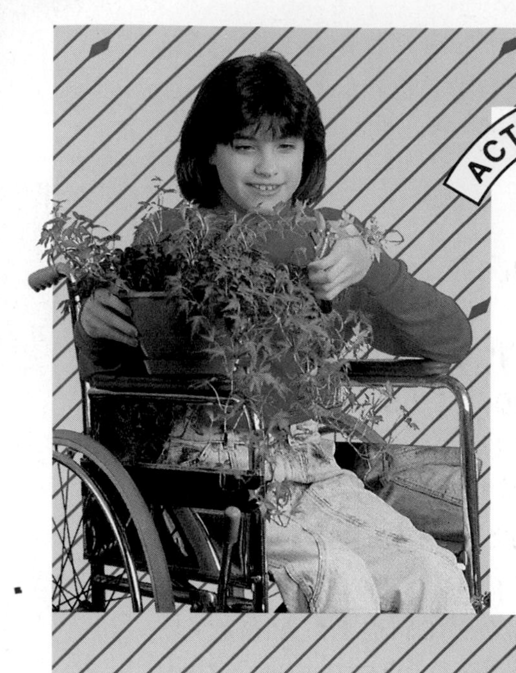

ACTIVITY You Can...

Grow a Plant Without Seeds

Cut off a small branch of ivy. Make the cut just below the place where a leaf joins the stem. Put the ivy into a jar of water. Observe the plant for four weeks. What happened to the ivy? Why might a person want to start a new plant from a plant part, instead of a seed?

Lesson Summary

- The parts of a flower include the sepals, petals, pistil, and stamens.
- Seeds, bulbs, rhizomes, tubers, and runners are plant parts that can produce new plants.
- The plant life cycle is the germination, growth, and seed formation of seed plants.

Lesson Review

1. Name two ways that pollination can take place.
2. Compare rhizomes, tubers, and runners.

★3. Explain the steps of the plant life cycle after a plant forms seeds inside fruit.

Use Application Activity on pages 359, 360.

I WANT TO KNOW ABOUT...

Plants in Space

Someday astronauts will be living in a space station somewhere between Earth and the moon. A problem that these astronauts will face within the enclosed space station is indoor air pollution. NASA scientists must find a way to keep the indoor air pollution levels low so that the air will be safe for the astronauts.

One part of the answer to the problem is very simple. Houseplants can be used to lower indoor pollution and keep the air safe to breathe. Scientists have found that some houseplants remove dangerous gases from the air. They think that the plants use these gases during photosynthesis. For example, the carbon dioxide exhaled by the astronauts is used by the plants. During photosynthesis, the carbon dioxide is broken down into oxygen and water. Astronauts can breathe the oxygen. The water given off by the plants will keep the air in the space station moist.

The NASA scientists plan to build a plant room in the space station. The room will be filled with plants that will filter the air in the space station.

CHAPTER REVIEW 5

Summary

Lesson 1
- Seeds have different sizes, shapes, and colors.
- A seed has stored food, an embryo, and a seed coat.
- Seeds need water and the right temperature to germinate.

Lesson 2
- Photosynthesis is the process by which plants make food.
- Plants use chlorophyll, carbon dioxide, and water to make food.

Lesson 3
- Flower parts include sepals, petals, pistil, and stamens.
- Seeds, bulbs, rhizomes, tubers, and runners produce plants.
- The plant life cycle is the germination, growth, and seed formation of seed plants.

Science Words

Fill in the blank with the correct word or words from the list.

embryo
germination
fruit
photosynthesis
chlorophyll
pistil
stamens
pollen
pollination
plant life cycle

1. The young growing plant inside the seed is the ___.
2. ___ are male flower parts.
3. ___ occurs when pollen grains move onto the pistil.
4. Stamens produce ___.
5. A chemical that traps energy from the sun is ___.
6. The ovary is in the ___.
7. Food is made during ___.
8. Seeds grow inside a(n) ___.
9. ___ begins when a plant embryo starts to grow.
10. The ___ is the germination, growth, and seed formation of seed plants.

Questions

Recalling Ideas

Correctly complete each of the following sentences.

1. Each seed contains
   (a) a tiny plant.
   (b) an outer skin.
   (c) stored food.
   (d) all of these
2. The outer skin of a seed is the
   (a) embryo. (c) seed coat.
   (b) pistil. (d) stamen.
3. The embryo of a seed grows into a young plant called a
   (a) seedling. (c) flower.
   (b) seed coat. (d) stamen.
4. Oxygen that animals breathe is made by
   (a) germination.
   (b) photosynthesis.
   (c) seeds.
   (d) pollination.
5. The outer parts of the flower that protect the flower before it opens are the
   (a) petals. (c) sepals.
   (b) seeds. (d) stamens.

Understanding Ideas

Answer the following questions using complete sentences.

1. How are seeds different?
2. What are some ways in which seeds are alike?

3. When do seeds germinate?
4. Describe how plants make their food.
5. Name the parts of a flower.
6. What are some plant parts that can be used to make new plants?

Thinking Critically

Think about what you have learned in this chapter. Answer the following questions using complete sentences.

1. Describe the changes that would occur on Earth if all the plants suddenly died.
2. Scientists consider green peppers and cucumbers to be fruits. Agree or disagree with this statement and give your reasons.

UNIT REVIEW 1

Checking for Understanding

Write a short answer for each question or statement.

1. What are the five steps of the scientific method?
2. What is the main difference between invertebrates and vertebrates?
3. What are complex invertebrates?
4. What is the difference between a cold-blooded and a warm-blooded vertebrate?
5. Name two groups of plants.
6. Name the parts of a seed.
7. What is photosynthesis?
8. How did Dian Fossey help mountain gorillas in Africa?
9. Give one characteristic of each of the invertebrate groups.
10. Give examples from each of the major groups of arthropods.
11. Describe the skeleton of a vertebrate.
12. How do the different groups of vertebrates reproduce?
13. Why are roots important for the growth of seed plants?
14. Why are leaves important for the growth of a plant?
15. What are the stages or parts of the plant-life cycle?
16. In what three ways are all seeds alike?

Recalling Activities

Write a short paragraph for each question or statement.

1. How does gravity affect plant growth?
2. How can data be collected?
3. How are sponges alike and different?
4. What are the characteristics of some complex invertebrates?
5. How does temperature change affect fish?
6. How do bones differ?
7. What are some different kinds of seed plants?
8. How does water move to leaves?
9. What are the conditions for seed germination?
10. How can you show that photosynthesis produces starch?

SCIENCE FAIR
Project Ideas
1. Show how seeds germinate under different conditions.
2. Investigate regeneration in simple animals. Make a diagram that will help to demonstrate this process.
3. Construct a plant maze to see how much a plant will move to get light. Make a maze in a cardboard box with a lid. Make an opening at one end of the box and put a vine plant inside the maze opposite the hole. Put the box where the hole gets light and water the plant.
MONTEREY BAY AQUARIUM
Books to Read
Discovering Worms by Jennifer Coldrey, Watts, 1986.
How Plants Grow by Steve Parker, Watts, Danbury, 1985.
Tiger Lillies and Other Beastly Plants by Elizabeth Ring, Walker and Company, NY, 1984.
Science Can Be Fun by Keith Wicks, Lerner Publications, Minneapolis, MN, 1987.
Plant Facts and Fancies by Sylvia Woods, Faber and Faber, Winchester, MA, 1985.

Earth Science

UNIT 2

Waves of the sea
make the sound of thunder
when they break against rocks
and somersault under.

Waves of the sea
make the sound of laughter
when they run down the beach
and birds run after.

"Waves of the Sea"
Aileen Fisher

SCIENCE
In Your World

Members of Our Solar System

A warm fire is great on a chilly day. But have you ever stepped back from the fire to keep from getting too hot?

The planets are similar. They feel the sun's heat as they circle around it. Some are close to the sun, and others are far away.

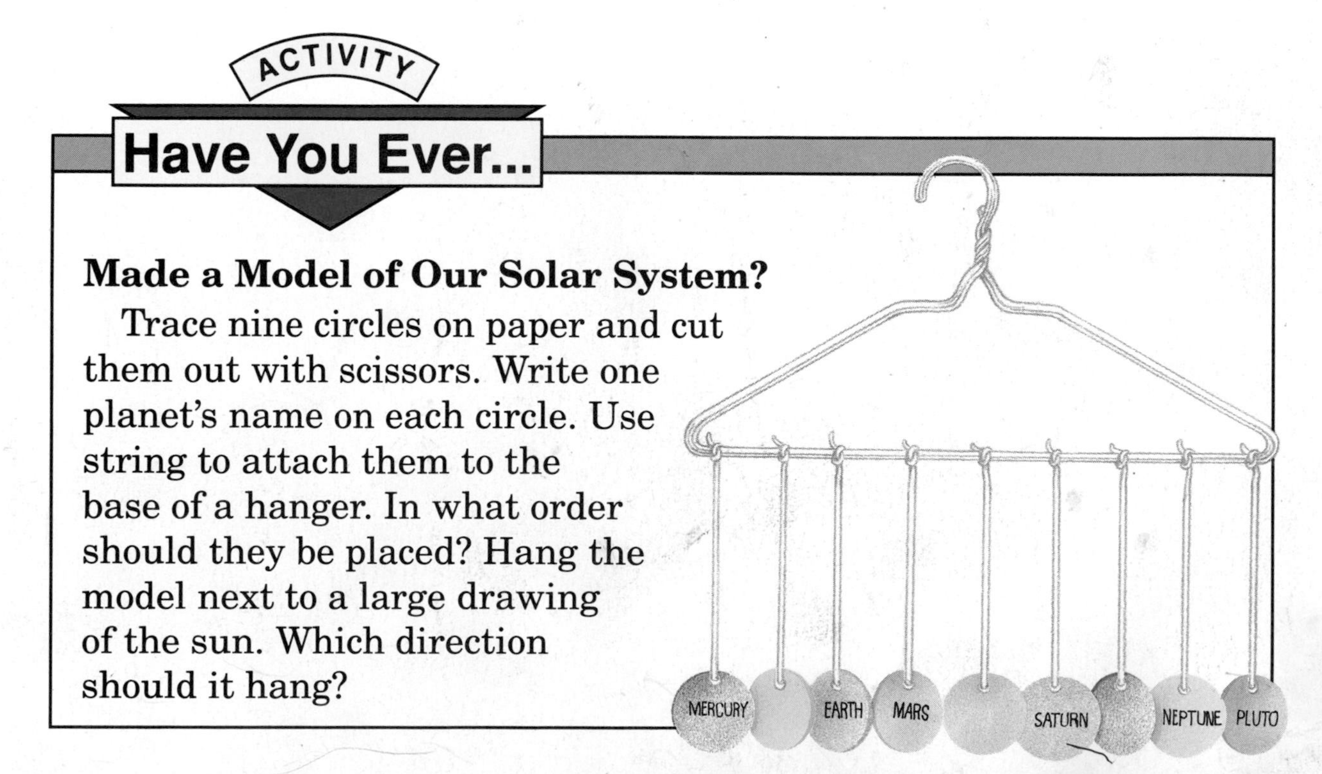

Have You Ever...

Made a Model of Our Solar System?

Trace nine circles on paper and cut them out with scissors. Write one planet's name on each circle. Use string to attach them to the base of a hanger. In what order should they be placed? Hang the model next to a large drawing of the sun. Which direction should it hang?

Planets and Orbits

LESSON 1 GOALS
You will learn
- what our solar system is.
- that the force of gravity causes each planet to travel in a certain path.
- what causes Earth's seasons.

What can you see if you look up into the clear night sky? You may be able to see lots of stars, some of them in patterns. Depending on what day of the month it is, you may be able to see the moon.

Early astronomers observed objects in the night sky and wondered what they were. Some of the objects they observed appeared to keep the same positions in the sky night after night. Other objects slowly but visibly changed their positions. Early astronomers called these moving objects "planets." The word meant wanderer. **Planet** is still used today to describe a large space object that revolves around the sun.

Our solar system

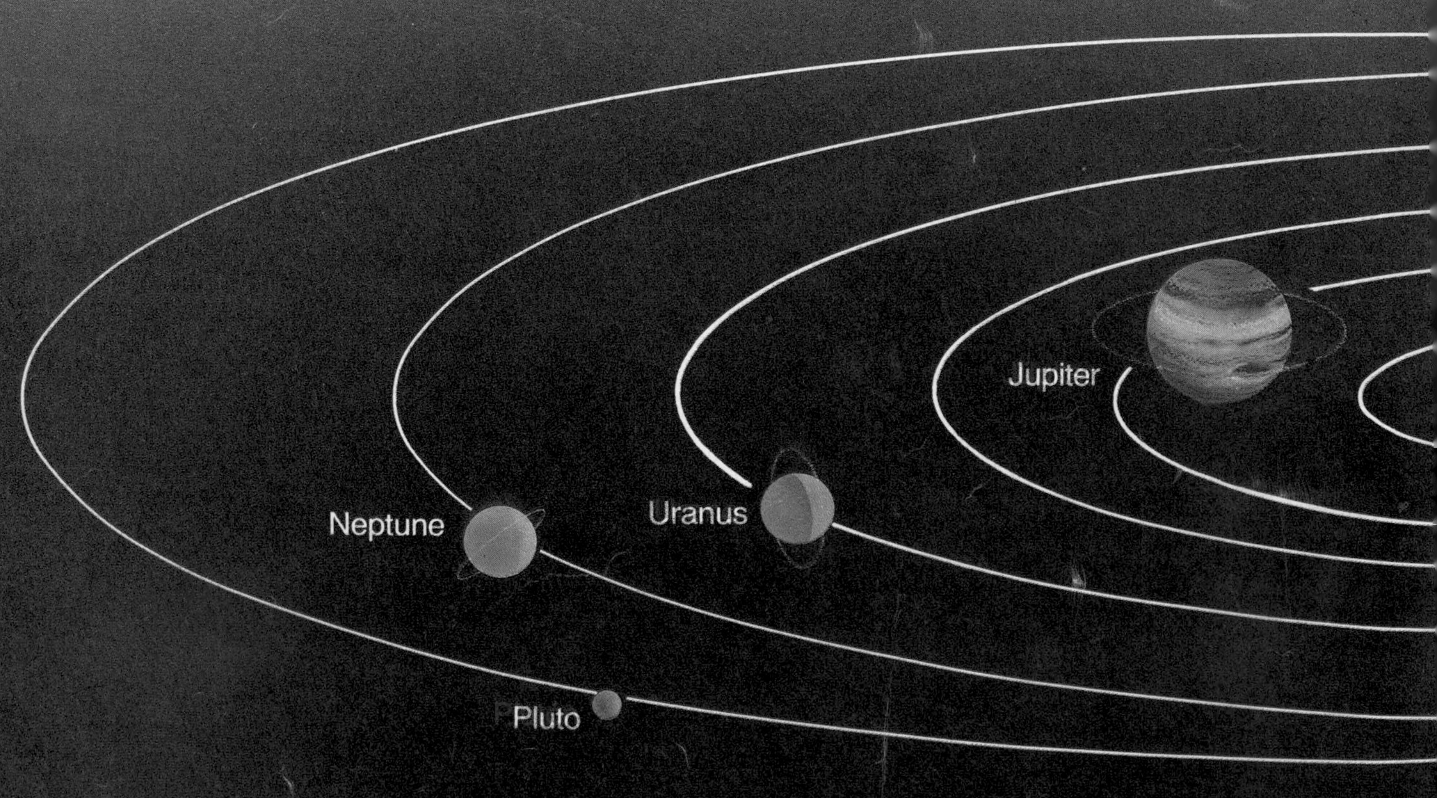

At first, astronomers thought that seven planets, including the moon and the sun, revolved around Earth. Today, we know that these space objects are part of a larger system. We know there are nine planets, including Earth, that revolve around the sun. The sun and all the space objects that travel around it make up our **solar system.**

Planets revolve around the sun in paths called **orbits.** The planets stay in their orbits because of gravity. **Gravity** is a force that causes objects to be attracted to one another. If there were no force of gravity, the planets would move in straight lines in space. But, because there is the pull of gravity between the sun and the planets, they revolve around the sun in orbits.

What keeps planets in orbit around the sun?

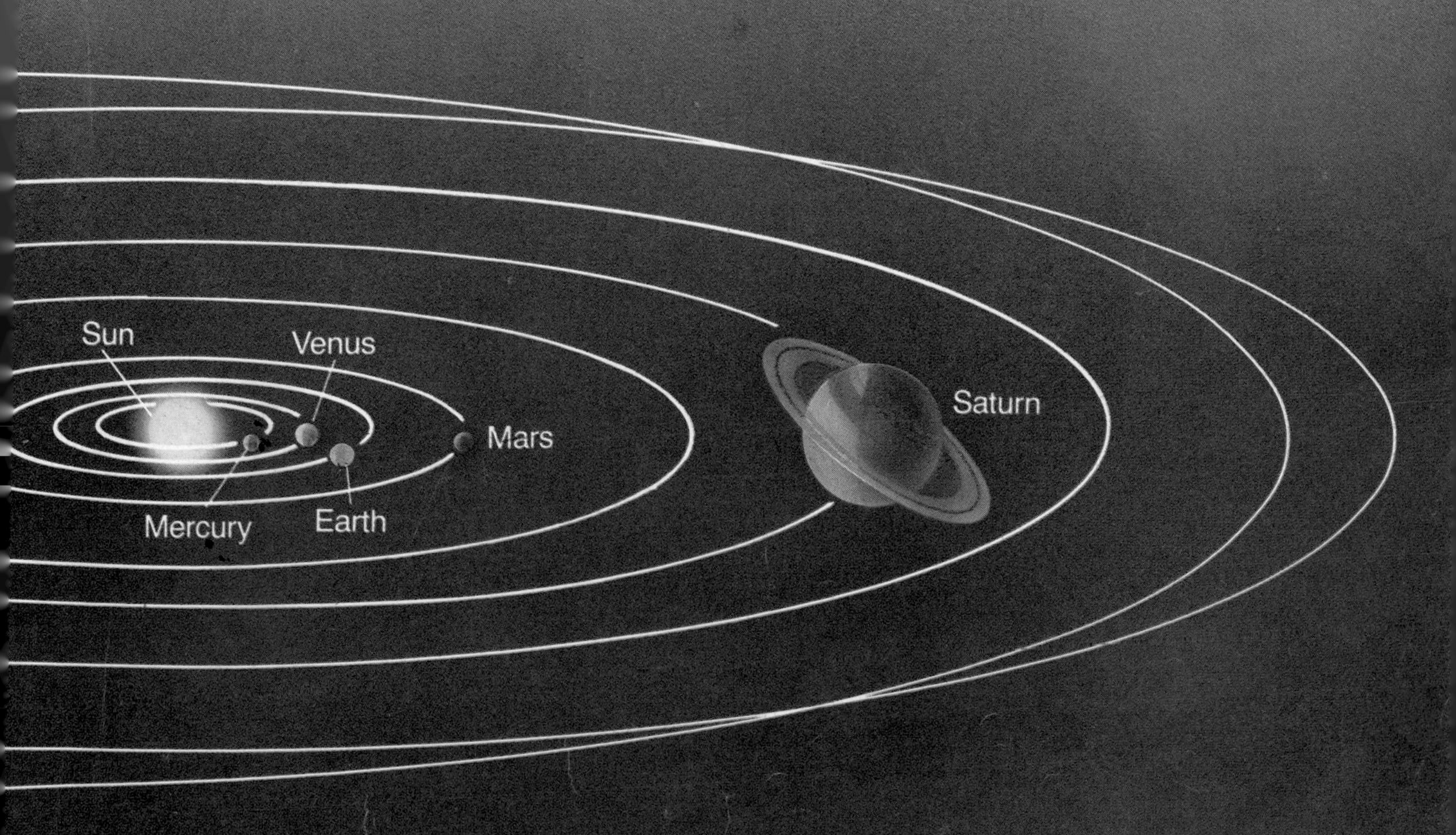

At first, scientists thought each planet followed a circular path around the sun. Now, however, we know that a planet revolves around the sun in an oval-shaped orbit called an **ellipse** (ih LIHPS). Look at the picture. You can see that the sun is not located in the center of a planet's orbit. So at times the planet is closer to the sun than at other times.

A planet's orbit

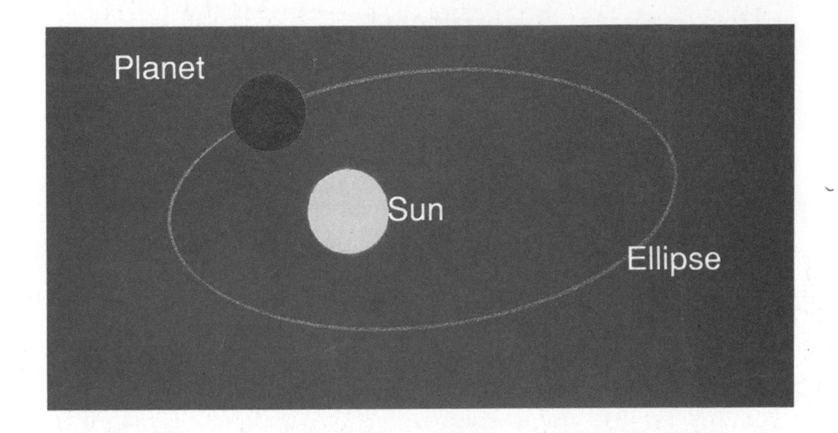

A planet speeds up when it is closer to the sun and slows down when it is farther away from the sun. A planet whose ellipse is more round travels almost the same speed throughout its orbit. Which planet, in the picture below, travels at about the same speed throughout its orbit?

The orbits of Mercury and Jupiter

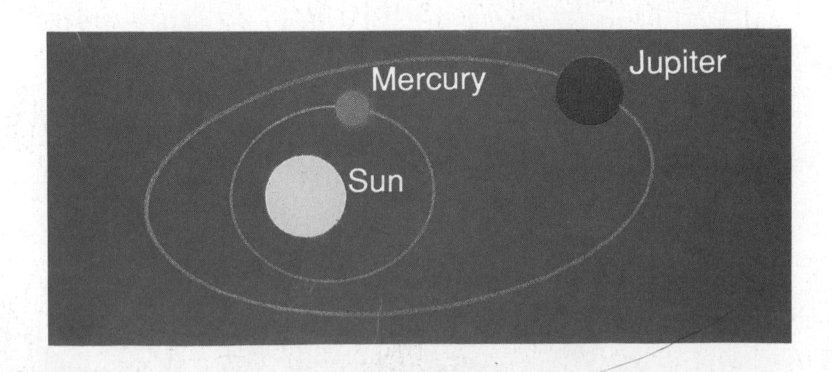

As planets revolve around the sun, they also rotate like a spinning basketball. The planets spin on an imaginary line through their centers called an axis. The ends of this axis are a planet's poles. What are Earth's poles called? Each planet spins at a different rate. Earth rotates on its axis once every 24 hours. One rotation is one day.

Each planet rotates on its axis.

The planets don't spin on a straight up-and-down axis like a basketball spinning on a finger does. The planets are all tilted as they spin. The amount of tilt varies for each planet. The tilt of Earth causes different amounts of sunlight to fall on certain places at different times of the year. In this way, the tilt of Earth's axis causes seasons.

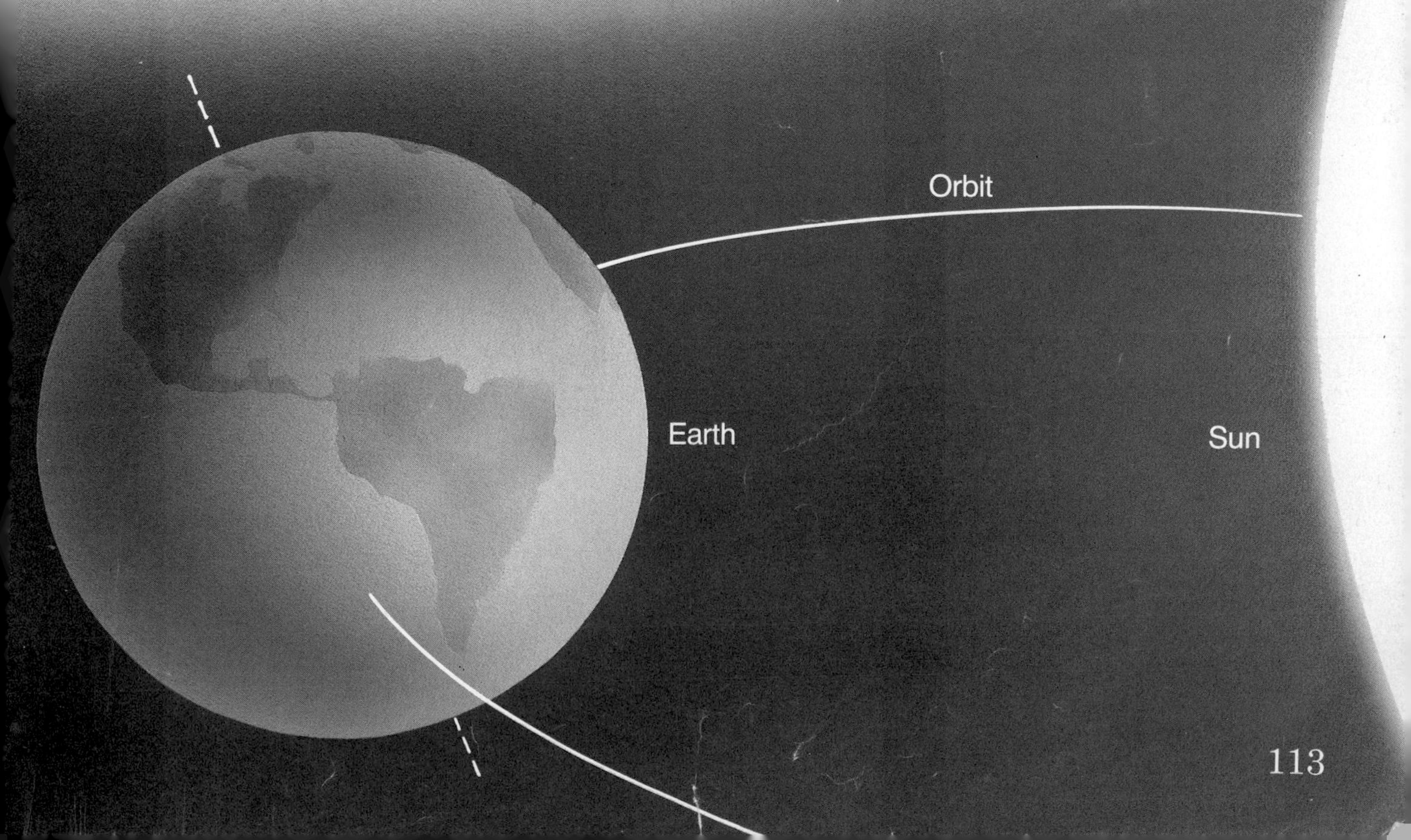

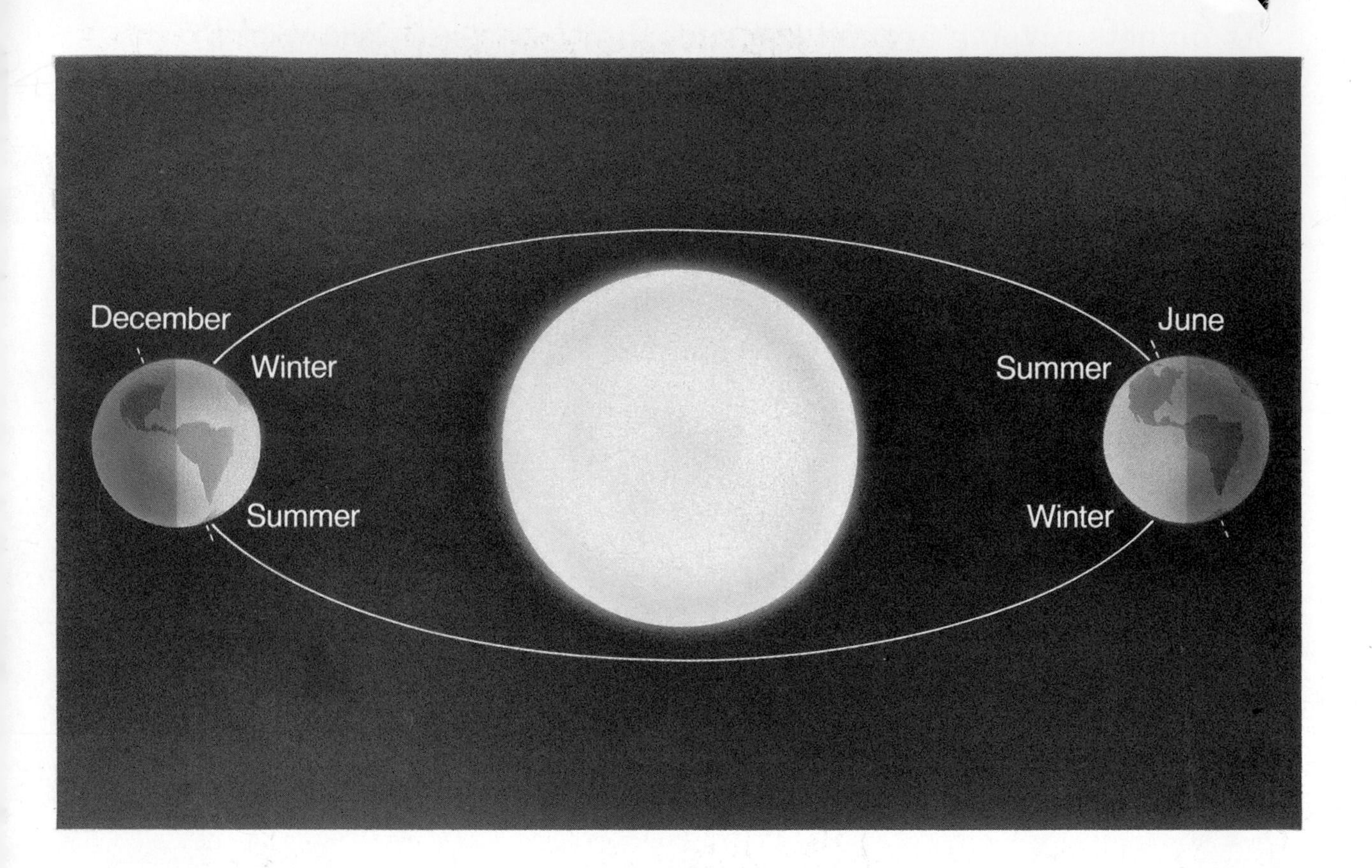

Earth's tilted axis causes seasons.

Look at the picture and notice that part of Earth is tilted toward the sun. When the top half of Earth is tilted toward the sun, it has summer. While the top half is tilted toward the sun, the bottom half of Earth is tilted away from the sun and has winter. When Earth revolves around the sun to the other side of its orbit, the top half tilts away from the sun and the bottom half tilts toward it. What seasons occur when Earth is in this position?

Spring and fall occur when Earth is halfway between the opposite sides of its orbit. In the time it takes for Earth to make one complete revolution, specific places on Earth have four seasons.

June 21st in different places on Earth

Lesson Summary

- The solar system is our sun and all the space objects traveling around it.
- The force of gravity keeps each planet in an orbit around the sun. Planets travel around the sun in orbits called ellipses.
- The tilt of Earth causes seasons.

Lesson Review

1. What does the word *planet* mean?
2. How would you describe an ellipse?
★**3.** How would the seasons be affected if Earth were not tilted on its axis?

Use Application Activity on pages 361, 362.

Inner Planets

LESSON 2 GOALS
You will learn
- which four planets are the closest to the sun.
- methods used to study space objects.
- how characteristics of the inner planets compare.

Imagine your class will blast off in a spacecraft to tour our solar system. On this tour you will visit the **inner planets**—Mercury, Venus, Earth, and Mars. They are called the inner planets because they are closest to the sun. They also are all rocky and are small in size when compared to other planets. What other things would you like to know about these planets? Keep your questions in mind as you prepare for your visit.

Space probe

More and more, scientists are using a variety of methods to study the planets. Even the early scientists used telescopes to view the planets. Today's scientists use telescopes, radar, and **space probes** to study space objects. They send space probes into space to gather information and send it back to Earth. You will want to learn all you can about what the space probes have uncovered. You can use this information to learn more about the planets.

Mercury

If you begin your tour with the planet closest to the sun, you will visit Mercury. Your visit to Mercury will be very short because it is so hot during the day and so cold at night. During the day, the temperature on Mercury can become as hot as 400°C. That's about twice as hot as an oven would be if you were to bake a cake. But at night, the temperature cools to −150°C. Mercury has a thin atmosphere because of its low surface gravity and high surface temperature. Sodium and helium have been found in Mercury's atmosphere. You may be familiar with helium. We use it to fill balloons.

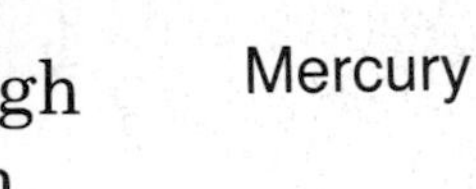

Mercury

You can look at photos taken by the space probe *Mariner 10* to see what Mercury's surface is like. Mercury looks a lot like our moon. It is covered with craters and has an unusual pattern of ridges and cracks. Most astronomers believe that the pattern of ridges happened when Mercury cooled and shrank shortly after it formed.

Venus

Why might Venus often be called Earth's twin?

As you leave Mercury, the next planet on your tour is Venus. You may already know that Venus is one of the brightest objects in the night sky. It is nearly the same size as Earth. In fact, Venus is often called Earth's twin.

Both radar and space probes have given us new information about Venus. At one time, people thought Venus might have water and plant life, but we now know that Venus has a very unpleasant environment. Unlike Mercury, Venus' atmosphere is very dense. You can't see its surface because it is hidden by thick clouds of sulfuric acid. But like Mercury, Venus' surface temperature is very hot. Do you think plants and animals could live on Venus? Why do you think so?

Venus

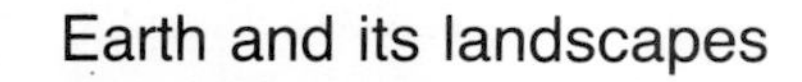

Earth and its landscapes

Earth

The next planet on your tour is Earth, the planet you live on. On your tour, you could see how it looks from space. What do you think your view of Earth would be like from space? Probably you could see some of the continents, clouds, and large bodies of water. A close-up view may show Earth's mountains, canyons, deserts, and craters.

Earth is the only planet in our solar system that has life. It has water on its surface and in its atmosphere. How does Earth's atmosphere compare to Mercury's or Venus' atmosphere? What important gas would you find in Earth's atmosphere?

A **satellite** is an object that revolves around another object. One satellite you might want to view is Earth's natural satellite—the moon.

SCIENCE AND . . .
Math

The planet Mars is about 141,500,000 miles from the sun. What is the value of the 5 in this number?

A. Hundreds
B. Thousands
C. Hundred thousands
D. Millions

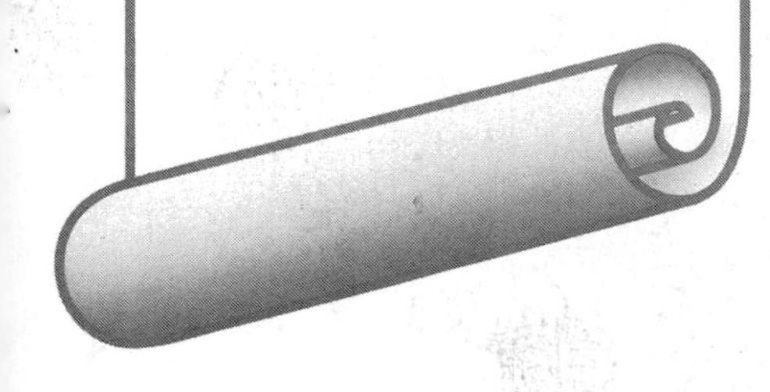

Mars

Mars is the last stop on your tour of the inner planets of our solar system. Just as we noticed Earth's natural satellite, we should also see that Mars has two moons. As you get closer to Mars, you will see that it has a reddish color on its surface. The *Viking* space probes, recently sent, discovered that Mars has huge canyons and a giant volcano bigger than any on Earth!

What do you think Mars' atmosphere will be like? Mars has a thin atmosphere that is mainly carbon dioxide and a small amount of water vapor. Thin ice crystal clouds form on Mars in the afternoon. Mars has little or no oxygen in its atmosphere. What do you think your chances would be of meeting a Martian or finding any other living things here?

Be careful where you land your spacecraft because Mars sometimes has strong winds. The wind blows soil into the atmosphere, causing large dust storms. These large dust storms cause changes in seasons on Mars.

Mars and its surface features

Now that you've toured the inner planets of our solar system, what similarities did you notice among some of the planets? What differences did you notice? Which planet, besides Earth, would you choose if you could make a return trip to study the planet in more detail? Why would you choose that planet?

Lesson Summary

- Mercury, Venus, Earth, and Mars are called inner planets.
- Telescopes, radar, and space probes are used to study space objects.
- Surface features, temperatures, and atmospheres of the inner planets can be compared.

Lesson Review

1. What is another name for a planet's natural satellites?
2. What gas is necessary for life to exist?

★3. Which planet takes the least time to travel around the sun?

Outer Planets

LESSON 3 GOALS
You will learn
- which five planets are the farthest from the sun.
- how characteristics of the outer planets compare.

In Lesson 2, you learned about the inner planets—Mercury, Venus, Earth, and Mars. These planets are small rocky bodies that are located close to the sun. You learned some interesting things about the inner planets. The next part of the tour of our solar system will take you to the **outer planets**—Jupiter, Saturn, Uranus, Neptune, and Pluto. These planets are farther from the sun. The first of these planets are called gas giants. The farthest planet from the sun is Pluto. It has many features like those of the inner planets.

Jupiter

The first planet on this tour, the fifth from the sun, is Jupiter. It is the largest of all the planets. When you get your first look at Jupiter, you will notice it is very large and very colorful. Its diameter is 11 times larger than Earth's. Its mass, or the amount of material that makes up Jupiter, is 300 times greater than Earth's mass.

Jupiter has a thin ring around it and at least 16 natural satellites, or moons. Jupiter's moons revolve around it much like the planets of the solar system travel around the sun.

Astronomers believe Jupiter is made mostly of the gases hydrogen and helium. Like the other gas giants, it has little if any solid surface. Its upper atmosphere has brightly colored bands of gases that look like clouds.

What gases do astronomers believe Jupiter is made up of?

As you look more closely at Jupiter, you will surely notice Jupiter's Great Red Spot. This red spot is over three times the size of Earth! Scientists think the spot is a giant storm in Jupiter's atmosphere.

It takes Jupiter 12 Earth years to make one orbit around the sun. But a day on Jupiter lasts only about ten Earth hours. How does that compare with the length of a day on Earth?

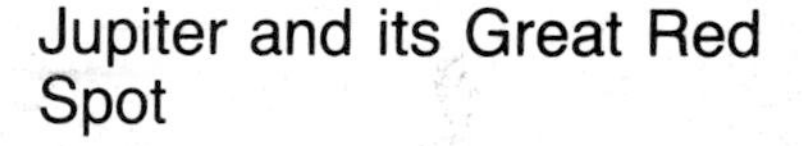

Jupiter and its Great Red Spot

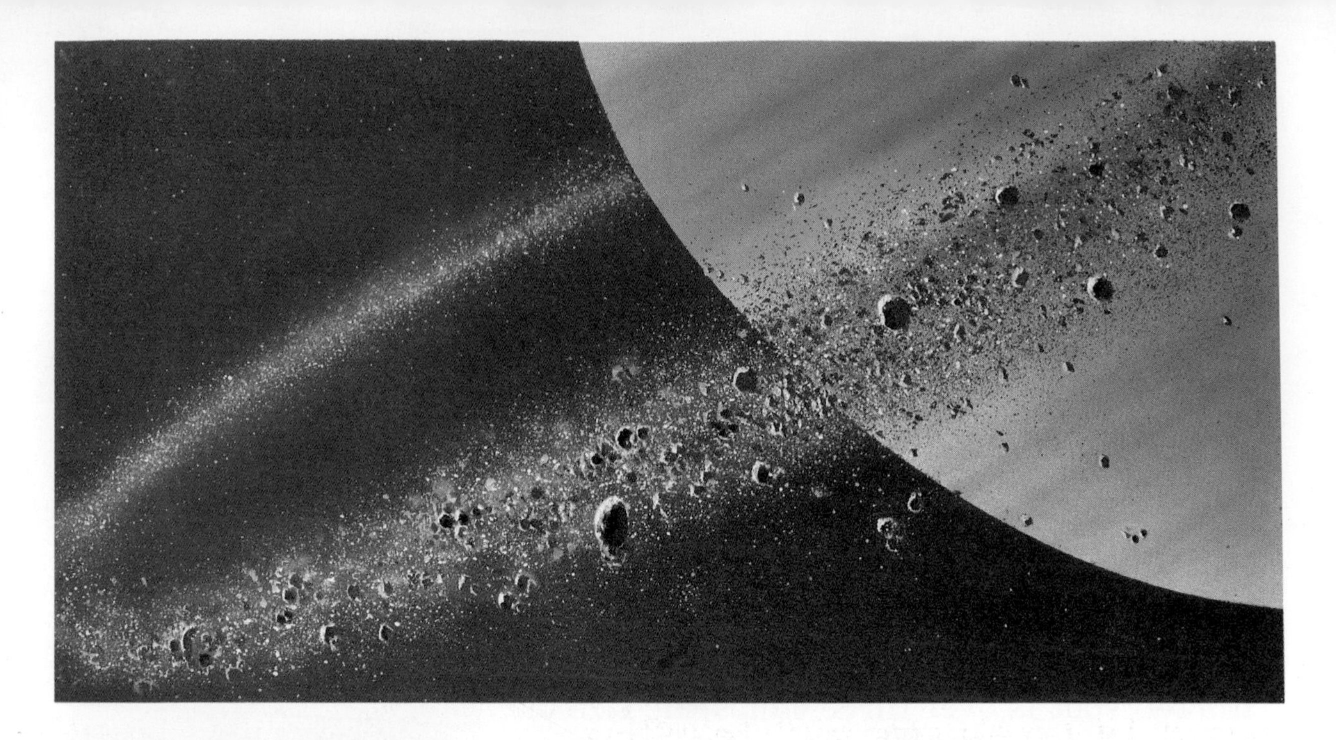

Saturn

Saturn

After leaving Jupiter, you will look next at Saturn, the second largest planet. Like Jupiter, it is made of hydrogen, helium, and other gases.

One of the things you will notice about Saturn is that it has many rings. You can see a few of Saturn's rings if you look at it through a telescope on Earth.

But to get a better look at the rings, scientists used the space probe *Voyager 2* to take close-up pictures. These pictures show that Saturn has over 1,000 rings! The rings are made of millions of small particles of ice and rock that orbit the planet.

From your spacecraft, you can observe that Saturn rotates once in just over ten Earth hours. You can see that Saturn has over 20 moons. Because of its distance from the sun, one year on Saturn is about 29½ Earth years long.

Uranus

The next stop on your tour is Uranus. As you approach Uranus in your spacecraft, you will first see its moons. Uranus has at least 15 moons orbiting it. Pictures received from *Voyager 2* also prepare us to see the rings around Uranus.

Like the other outer planets you've seen so far, Uranus is made of gases. It has very few markings and looks like a large, smooth blue-green ball.

Unlike other planets, Uranus is tilted on its side as it revolves around the sun. Thus, part of the time it appears to be rolling along its orbit. At these times, one of the poles points directly toward the sun. The other is opposite the sun.

Uranus

Neptune

Neptune is the next planet you will visit. It looks much like Uranus but it is pale blue in color. It too is made of hydrogen, helium, and other gases.

Neptune

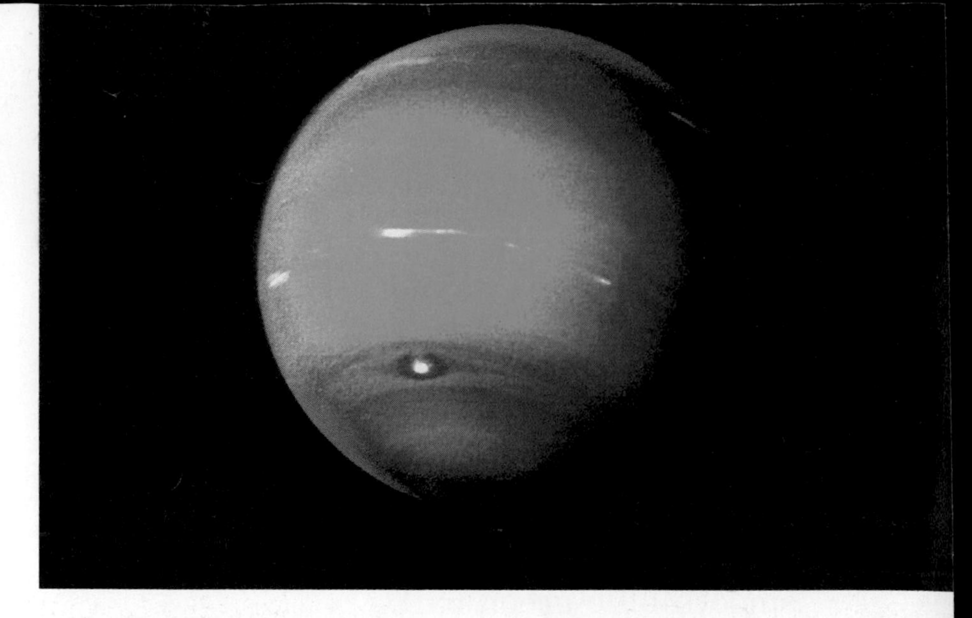

You'll notice that Neptune's orbit is large. A Neptune year is equal to 165 years on Earth. A day lasts 16 Earth hours.

Voyager 2 has also explored this planet. It sent pictures to Earth that showed Neptune has violent weather with extremely high winds. Neptune has at least eight moons and four complete rings around it.

Pluto

Pluto is the last planet on your tour of our solar system. It is called the ninth planet, but it is not always the ninth planet from the sun. During part of its orbit, it travels inside the orbit of Neptune. When it's outside of Neptune's orbit, it is the farthest planet from the sun.

You will notice that Pluto is more like the inner planets because it has a solid surface. It appears to be made of ice and gases. Because of its great distance from the sun, Pluto is a cold, dark place. It has one moon. Pluto takes 248 Earth years to complete one orbit.

Pluto

After taking a "quick tour" of the planets of our solar system, you will know more about each of their characteristics. Did you find the answers to some of your questions? As scientists learn more things, they usually think of more questions to ask. You too will probably think of more things you want to know about the planets of our solar system.

Would You Believe?

On Pluto, the sun would appear no brighter than Venus appears to us.

Lesson Summary

- The outer planets are Jupiter, Saturn, Uranus, Neptune, and Pluto.
- Unique features, orbits, and the number of moons of the outer planets can be compared. Jupiter, Saturn, Uranus, and Neptune are called the gas giants. Pluto, the farthest planet from the sun, has a solid surface like the inner planets.

Lesson Review

1. What planet is known for its Great Red Spot?
2. Why is Pluto not always the ninth planet from the sun?

★3. What causes one planet to have a shorter day than another?

The Moon

LESSON 4 GOALS
You will learn
- how the force of Earth's gravity affects Earth's atmosphere and the moon.
- how the moon moves around Earth.
- that the moon goes through phases.

Do you have some questions about the moon? If you do, you aren't alone. People have been interested in the moon for thousands of years.

For years, people studied the moon from a great distance. Then, between 1969 and 1972, scientists gathered first-hand information when they sent the *Apollo* spacecraft to the moon. The *Apollo* astronauts actually landed on the moon to do their work. They brought samples of rock and soil back to Earth for study. They also took pictures to record the location of different types of rocks.

Why is it important to compare the moon to Earth? The **moon** is Earth's natural satellite. As we learn more about the moon, we can learn more about Earth.

Man on the moon

Scientists have learned a lot about gravity by comparing Earth and the moon. We know that gravity pulls you back to Earth if you jump in the air. Earth has more gravity than the moon because it has more mass than the moon. Earth's mass is three and a half times that of the moon. Earth's gravity is six times that of the moon. On Earth, you weigh six times what you would on the moon because of this difference in gravity. If you weigh 60 pounds on Earth, how much would you weigh on the moon?

How does Earth's mass compare to the moon's mass?

The moon's gravity affects weight.

Did you realize that it is because of gravity that Earth has an atmosphere? The strength of Earth's gravity keeps the atmosphere from moving away. But the moon's gravity is too weak to hold an atmosphere.

It is gravity that causes the moon to revolve around Earth. Since the moon is smaller, Earth's gravity keeps the moon from moving away. As a result, the moon makes one revolution around Earth every 27 days. As the moon orbits Earth, it also rotates or spins on its axis. The moon makes one rotation during the same amount of time that it revolves once around Earth. How long is a moon day?

The moon does not make its own light. Moonlight is really reflected light from the sun. The light bounces off the moon and shines on Earth.

Have you noticed that the moon looks different at different times in a month? As the moon revolves around Earth, the moon's shape seems to change. We call these changes **moon phases.**

Moon phases

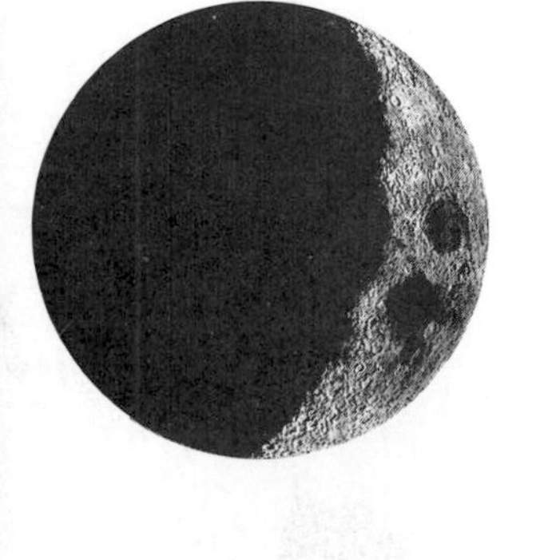

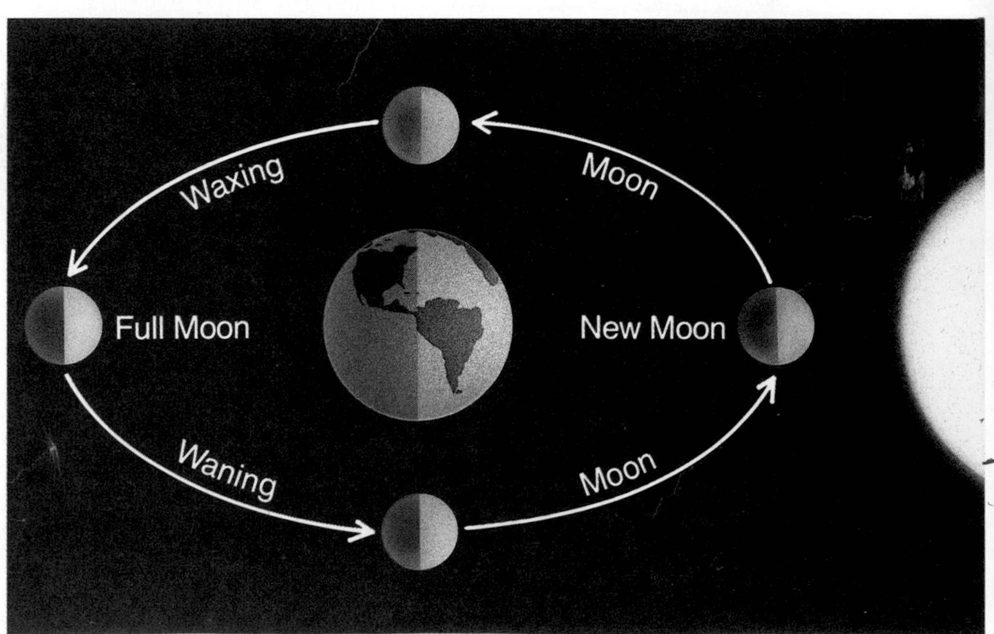

ACTIVITY

You Can...

Model How the Moon Moves

Make believe you are the moon and your friend is Earth. Walk counterclockwise around your friend. Keep one shoulder always pointed toward your friend. How many walls do you, the "moon," face as you revolve around "Earth"? Now stand in one place and rotate slowly. How many walls do you face as you rotate?

A new moon happens when the moon is between Earth and the sun. During this phase, all of the lighted side of the moon faces away from Earth, so we can't see the moon at all. Each night as the moon moves in its orbit around Earth, more of the lighted side can be seen. People say the moon is waxing when more of its lighted side shows each night.

Halfway through the moon's phase cycle, we see a full moon. This happens when Earth is between the moon and the sun. At this time, we can see all of the lighted side of the moon. After a full moon, as the moon continues in its orbit around Earth, we see less of the moon each night. The moon is waning during this phase.

Lesson Summary

- Earth's gravity holds an atmosphere and causes the moon to revolve around Earth.
- The moon makes one rotation during the same amount of time that it revolves once around Earth. The moon makes one revolution around Earth every 27 days.
- During a new moon, all of the lighted side of the moon faces away from Earth. The moon is waxing when more of its lighted side shows each night. When all of the lighted side can be seen, it is a full moon. The moon is waning when less of its lighted side shows each night.

Lesson Review

1. How much greater is Earth's gravity than that of the moon?
2. Which phase is halfway through the moon's cycle?

★3. If the moon does not make its own light, why does the moon shine so brightly on Earth?

Use Application Activity on pages 363, 364.

What are moon phases?

What you need

flashlight
globe
white softball
pencil and paper

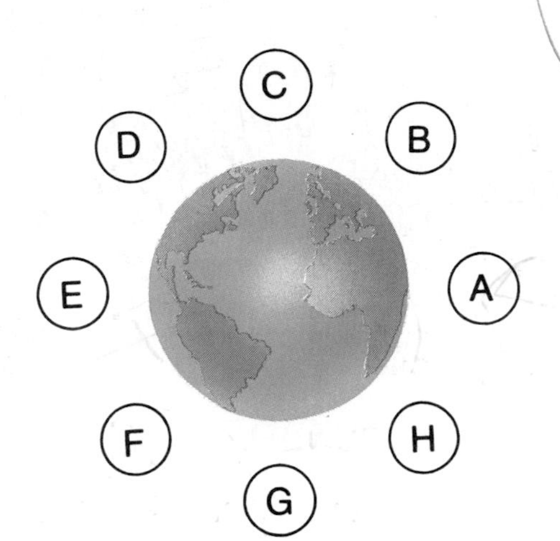

What to do

1. Make a drawing like the one shown.
2. In a darkened room, hold the flashlight. Have a partner hold the globe so your country faces the flashlight.
3. Have a third student hold the ball between the globe and flashlight.
4. Turn on the flashlight. Beginning at the letter A of your drawing, shade in how much of the lighted side of the ball you would see from the globe.
5. Move the ball counterclockwise to each place on the drawing. Draw how much of the lighted side of the ball you would see from the globe for each location.

What did you learn?

1. How much of the ball is always lighted?
2. How does the shape of the lighted part of the ball you can see from the globe, change as it moves around the globe?

Using what you learned

1. What do the flashlight, ball, and globe represent?
2. What happens after the moon completes a cycle of phases?

Other Members of Our Solar System

LESSON 5 GOALS
You will learn
- that other smaller space objects are members of our solar system.
- the characteristics of asteroids, comets, and meteoroids.

In earlier lessons, you learned about the planets and their satellites that make up our solar system. But there are other objects that orbit the sun, too. Asteroids, comets, and meteoroids are also members of the solar system. Although some are too small to be seen from Earth, others can be seen without a telescope.

Asteroids

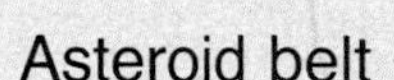
Asteroid belt

Asteroids are rocky space objects that orbit the sun. They are made of rock, metal, or minerals. There are thousands of asteroids in a wide band called the asteroid belt. The asteroid belt is located between the orbits of Mars and Jupiter. Asteroids take five Earth years to complete one orbit around the sun.

Asteroids can be small or up to several hundred kilometers in diameter. Some astronomers think that asteroids are leftover matter that never formed a planet.

Comets

Comets are space objects made of ice mixed with dust particles. They probably come from the far, outer edges of our solar system. They orbit the sun. We can see them only when they are close enough to be heated by the sun. It is then that comets give off light.

Comet Halley

You can think of a comet as being a dirty snowball. It has three parts—a nucleus, coma, and tail. The nucleus is made of ice, gases, and particles of rock. As a comet nears the sun, gases escape from the nucleus. A large, fuzzy, ball-shaped cloud, called the coma, is formed. The tail is present only when the coma is heated by the sun.

Comets have very large orbits. They take a long time to orbit the sun. Some comets take many Earth years to complete one orbit. A person may see only two or three naked-eye comets in a lifetime.

Meteoroids

When are meteoroids called meteors?

Meteoroids (MEET ee uh royds) are small pieces of metal or rock that orbit the sun. There are many meteoroids in space. Many are too small to be seen from Earth. Sometimes their orbits bring them close to Earth. When they travel through Earth's atmosphere, they are called meteors. On a clear night, you may have seen a meteor streak across the sky, leaving a trail of light. What is another name that is sometimes used for a meteor?

Most meteors burn up in the atmosphere 50 to 100 kilometers above Earth's surface. Some reach Earth's surface as small particles of metal, rock, or dust. A few very large ones do strike Earth's surface. A meteor that strikes Earth's surface is a meteorite (MEET ee uh rite).

Barringer Crater and meteor trail

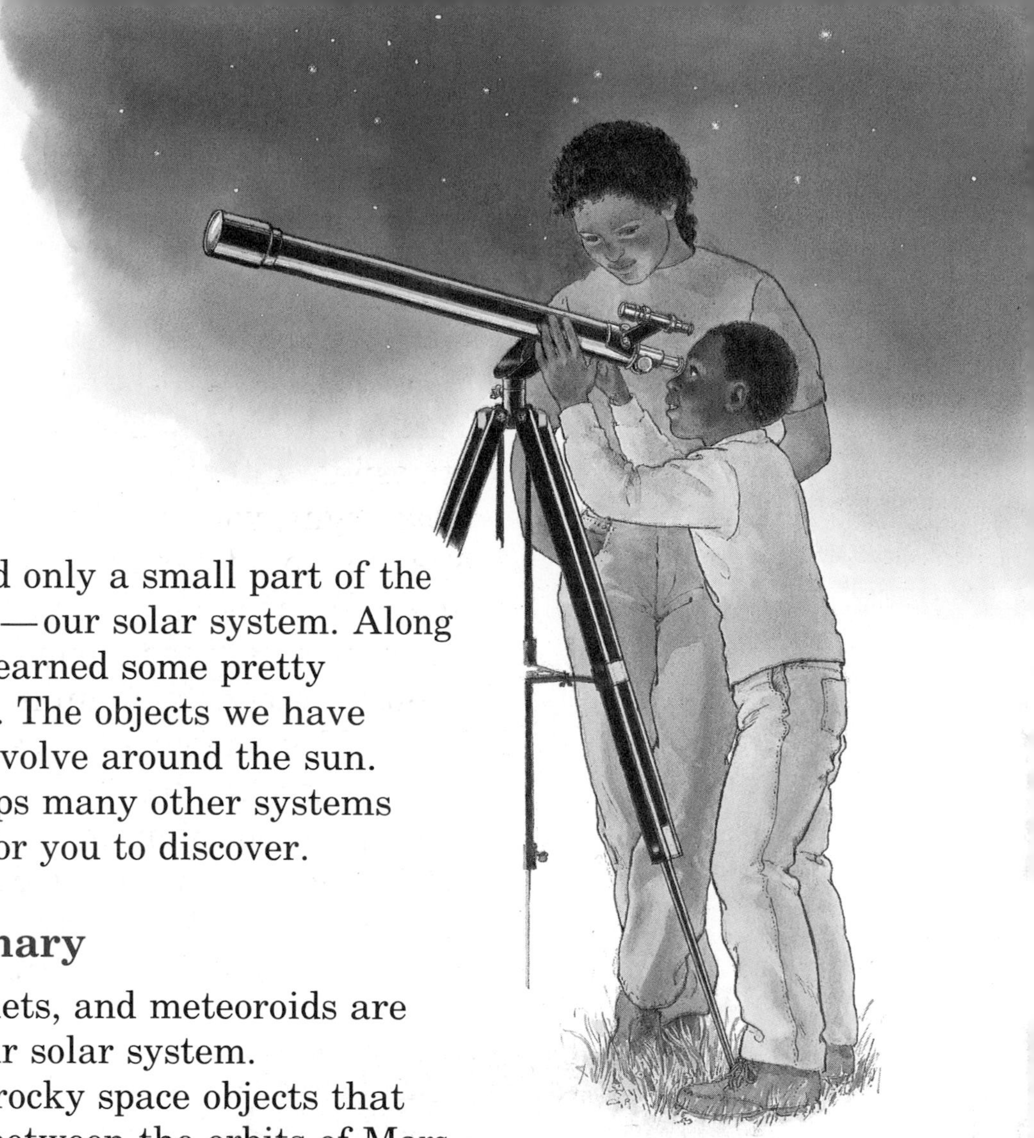

We've explored only a small part of the sky that you see—our solar system. Along the way, we've learned some pretty interesting facts. The objects we have learned about revolve around the sun. There are perhaps many other systems similar to ours for you to discover.

Lesson Summary

- Asteroids, comets, and meteoroids are members of our solar system.
- Asteroids are rocky space objects that orbit the sun between the orbits of Mars and Jupiter. Comets are space objects made of ice mixed with dust particles. Meteoroids are small pieces of metal or rock that are scattered in different orbits around the sun.

Lesson Review

1. Between which two planets' orbits is the asteroid belt found?
2. Why might a person see only two or three comets in a lifetime?

★3. What would cause a meteor to strike Earth?

How can you compare sizes of planets?

What you need

construction paper
compass
metric ruler
scissors
pencil and paper

What to do

1. Use the compass. Draw a circle on the paper to represent each planet and the sun. Check the chart to make sure you have the correct size.
2. Cut out and label each planet.
3. Arrange the planets so they are in the correct order from the sun. Fill in columns 1 and 2 in the chart provided by the teacher.
4. Divide the planets into 2 groups according to size. One group is Jupiter-sized. One group is Earth-sized. Fill in column 3.

What did you learn?

1. Which planets are about the same size as Earth?
2. Which planets are large like Jupiter?
3. What statement can be made about a planet's size and its position from the sun? Are there any exceptions to this statement?

Using what you learned

1. How do the sizes of the planets compare to the size of the model sun?
2. How is your model different from, or similar to, our solar system?

Planet	Diameter
Mercury	.5 cm
Venus	1.2 cm
Earth	1.3 cm
Mars	.7 cm
Jupiter	14.6 cm
Saturn	12.2 cm
Uranus	5.2 cm
Neptune	5.0 cm
Pluto	.3 cm

I WANT TO KNOW ABOUT...

Graphs

A graph is a kind of picture that shows a connection between two facts. It shows how they are alike and how they are different. In the graphs below, the word *gravity* is explained by comparing each planet's gravity.

What does the graph on the left tell you about gravity? Compare the numbers on the left side with the names across the bottom.

Put your finger on the number .25. Look at the names at the bottom of the graph. Slide your finger beside .25 to the spot just above the name *Mercury*. A point was placed at a spot between .25 and .5 to show the amount of gravity that Mercury has compared to Earth's gravity.

- What is the amount of gravity that each planet has compared to Earth's gravity?

A line was drawn to connect the points on the graph. This line shows how the planets' gravities compare to Earth's gravity.

The graph on the right is a bar graph showing the gravity of each planet. By seeing how the bars vary in size, you can get an idea of how the gravities are different.

- Which planet has the greatest gravity?
- Which planet has the least?

Gravity of the Inner Planets

Amount of gravity

1
.75
.5
.25
0

0.38
0.88
1.0
0.38

Mercury
Venus
Earth
Mars

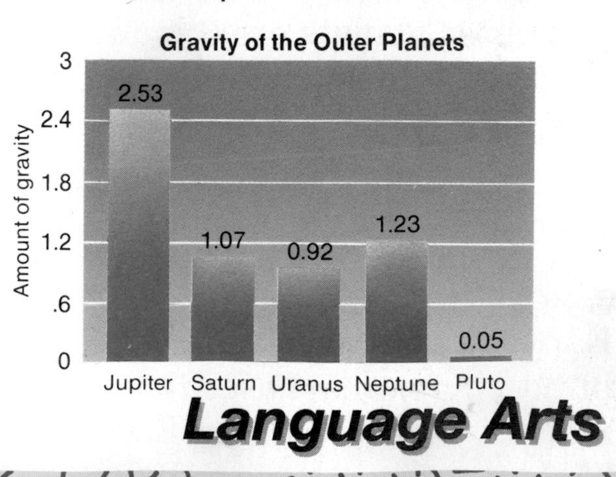

Language Arts

CHAPTER REVIEW 6

Summary

Lesson 1
- The solar system is our sun and space objects around it.
- The force of gravity keeps each planet in an orbit.
- Earth's tilt causes seasons.

Lesson 2
- Mercury, Venus, Earth, and Mars are the inner planets.
- Telescopes, radar, and space probes are used to study space.
- Characteristics of inner planets can be compared.

Lesson 3
- The outer planets are Jupiter, Saturn, Uranus, Neptune, and Pluto.
- Characteristics of outer planets can be compared.

Lesson 4
- Earth's gravity holds an atmosphere and causes the moon to revolve around Earth.
- The moon makes one rotation and one revolution around Earth every 27 days.
- During a new moon, all of the lighted side faces away from Earth. During a full moon, the lighted side is seen.

Lesson 5
- Asteroids, comets, and meteoroids are members of our solar system.
- Asteroids are rocky objects between Mars and Jupiter. Comets are ice and dust. Meteoroids are metal or rock.

Science Words

Fill in the blank with the correct word or words from the list.

planet	**ellipse**
solar system	**gravity**
orbits	**inner planets**
space probes	**outer planets**
satellite	**moon**
moon phases	**asteroids**
comets	**meteroids**

1. Rocky space objects that orbit the sun are ___.
2. Apparent changes in the moon's shape are its ___.
3. Space objects made of ice mixed with dust particles are called ___.
4. Planets revolve around the sun in paths called ___.
5. An object that revolves around another object is a(n) ___.

Questions

Recalling Ideas

Correctly complete each of the following sentences.

1. The half of Earth tilted away from the sun has
   (a) spring. (c) summer.
   (b) winter. (d) fall.
2. Space objects are studied by
   (a) telescopes. (c) radar.
   (b) probes. (d) all of these
3. Planets orbit the sun because of its
   (a) heat. (c) gravity.
   (b) brightness. (d) all of these
4. Our solar system contains planets, their satellites and
   (a) asteroids. (c) comets.
   (b) meteoroids. (d) all of these

Understanding Ideas

Answer the following questions using complete sentences.

1. Name the planets and describe each.
2. Describe two effects of Earth's gravity.

Thinking Critically

Think about what you have learned in this chapter. Answer the following questions using complete sentences.

1. What would your weight be on the moon?
2. If a moon year is the time it takes the moon to revolve around the sun, how many moon days are in a moon year?

SCIENCE
In Your World

CHAPTER 7

Our Star

The sun is only one of billions of other stars. What makes this huge yellow ball of hot, glowing gases so special? It is the closest star to Earth. Even though its size is bigger than a million earths, some other stars are actually much bigger and hotter. Why do you think they look much smaller and dimmer than they really are?

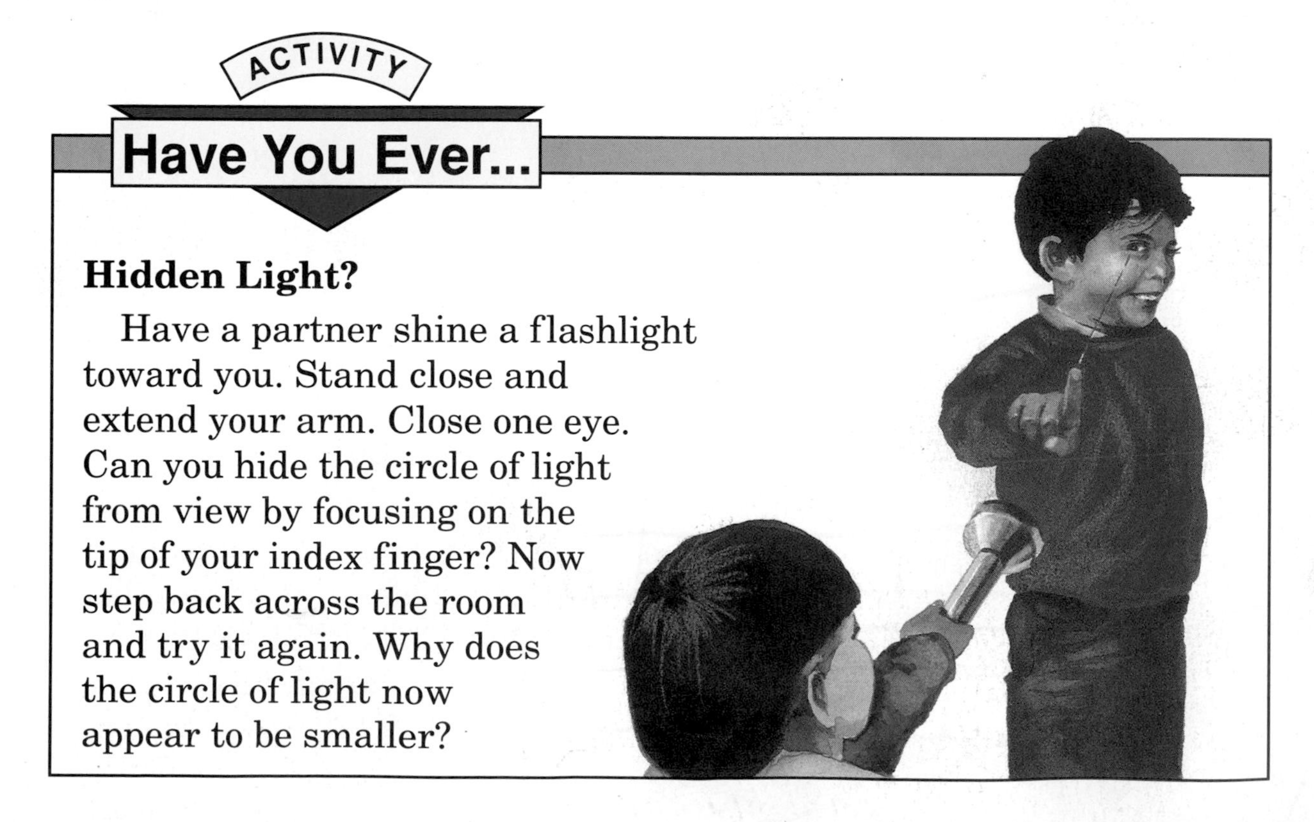

ACTIVITY

Have You Ever...

Hidden Light?

Have a partner shine a flashlight toward you. Stand close and extend your arm. Close one eye. Can you hide the circle of light from view by focusing on the tip of your index finger? Now step back across the room and try it again. Why does the circle of light now appear to be smaller?

Comparing Stars

LESSON 1 GOALS
You will learn
- that stars can be compared.
- how energy is produced in the sun's core.

"Twinkle, twinkle, little star. How I wonder what you are."

Over the years, people have wished on stars or wondered what they are. Without television, radio, or books, people in ancient times spent a lot of time looking at the stars. They saw the stars as patterns and imagined lines connecting the stars. Their patterns formed familiar shapes, much like connect-the-dot pictures you have made. One famous star pattern, the Big Dipper, looks like its name.

Our Sun as a Star

Not all the lights you see at night are stars. Some are planets, comets, and meteors that you learned about in Chapter 6. On the other hand, the only light you see during the day, the sun, is a star. The **sun** is the star that is closest to Earth. That is why it seems so much larger and brighter to us than the other stars we see at night.

The sun, like other **stars,** is a huge ball of glowing gases. It is the center of our solar system. The sun is so large that over one million Earths would fit inside the sun. Because planets are so much smaller than stars, no one knows if other stars have solar systems or not. The planets may just be too small to see. Astronomers say that our sun is an average-sized star.

How does our sun compare in size to other stars?

You may notice that stars are different colors. They can have shades of white, blue, yellow, orange, or red. Our sun is a yellow star. Its surface temperature is about 6,000°C. In its center, the sun is even hotter. Think about the temperature outside today. The sun is quite a bit hotter, isn't it? Stars that are white or blue are hotter than the sun. Orange and red stars are cooler.

The sun—our star

ACTIVITY

You Can...

See Colors in the Night Sky!

Go outside on a clear night. Use a pair of binoculars to study at least 15 different stars. Look carefully at their colors. How many look red, blue, yellow, or white? Did you observe any other colors? Make a table or bar graph showing how many stars of each color you observed. What could cause the stars to have different colors?

If you look at the stars, you notice that some are brighter than others. There are three reasons for this. First, stars give off different amounts of energy. The more energy that is given off, the brighter the star. Second, stars closer to Earth appear brighter than those farther away. Third, the size of the star affects how bright it appears. Large stars may appear brighter than small stars.

What is the average life span of a star?

Stars are not all the same age, either. The average life span of a star is about ten billion years. By watching other stars form and die, astronomers can tell that the sun is in the middle of its life. That would make it about five billion years old.

The sun, like other stars, began as a cloud of dust and a gas called hydrogen. The particles in the cloud were attracted to each other and the cloud got smaller. Heat and pressure increased in the cloud and it began to produce energy.

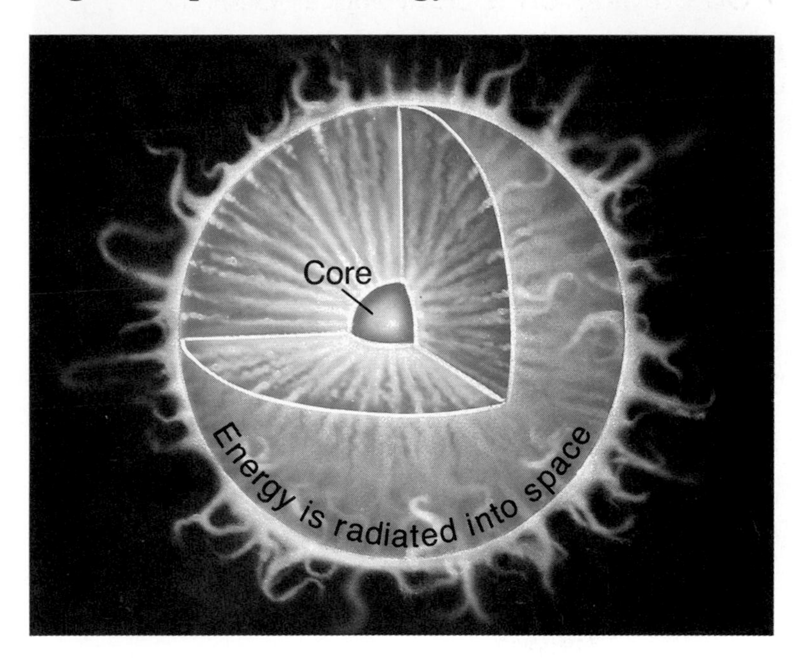

Fusion occurs in the sun's core.

Energy is produced when atoms of hydrogen join, or fuse. These atoms make a new gas, called helium. When helium is formed, energy is released. This process of joining atoms together is called **fusion** (FYEW zhun). The energy from fusion is released into space through the surface of the sun. The sun will continue to produce energy for another five billion years.

When the sun runs out of hydrogen fuel, it will get smaller. Then its outer layers will expand into what astronomers call a red giant. When the sun turns into a red giant, it will be much bigger than it is today. The red giant will collapse to form a small dense star called a white dwarf.

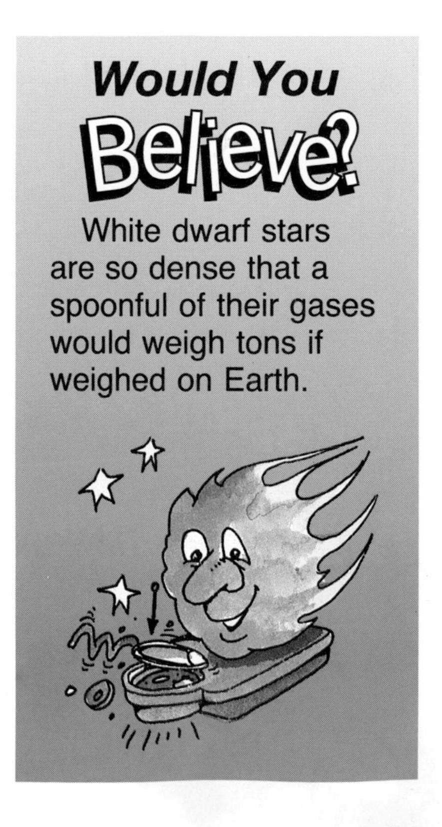

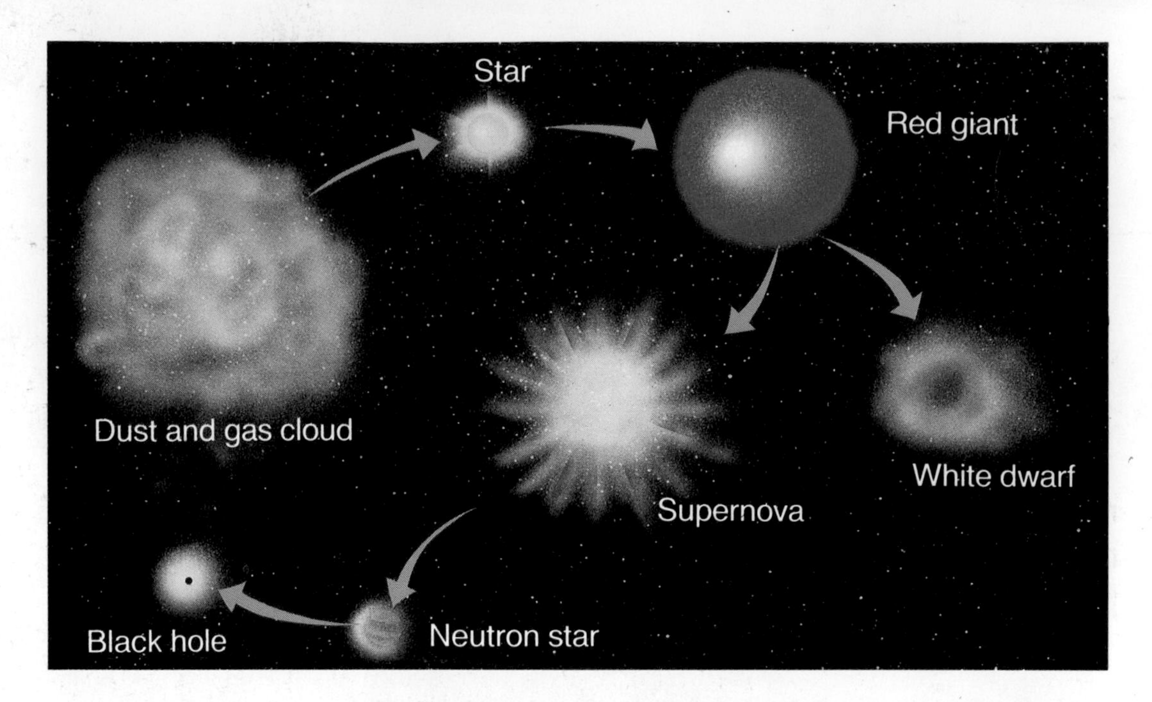

Life cycles of stars

Other stars have a similar cycle. But when a very large star becomes a **red giant** and collapses, the tremendous pressure causes it to explode. The exploding red giant is called a **supernova.**

The largest of supernovas may form **neutron** (NEW trahn) **stars** and possibly black holes. **Black holes** are stars with gravity so strong that even light can't escape. A black hole is the last possible stage in a star's life cycle.

Lesson Summary

- The sun is an average-sized yellow star that is about five billion years old.
- The sun's core produces energy when hydrogen atoms fuse to make helium.

Lesson Review

1. Why do stars appear different colors?

★2. Why do some stars appear brighter than others?

How do distance and size affect how bright a light looks?

What you need

3 flashlights (2 same-sized, 1 larger)
meter tape
movie screen
pencil and paper

What to do

1. In a darkened room, stand with a partner 1 meter away from a screen.
2. Each of you shine the same-sized flashlight on the screen.
3. Compare the brightness of the beams and record your observations.
4. Have your partner move back 1 meter.
5. Shine both flashlights. Record your observations.
6. Have your partner return to the first position but use the larger flashlight.
7. Shine both flashlights, compare the brightness, and record your observations.

What did you learn?

1. At 1 meter, how did the beams from the same-sized flashlights compare?
2. What happened to the size and brightness of the beam when your partner moved back?
3. How did the brightness of the beams from the two different-sized flashlights compare?

Using what you learned

1. Why is it important to use fresh batteries in both flashlights?
2. Based on what you have learned, give two reasons why one star might appear brighter than another.

The Sun and Earth

LESSON 2 GOALS
You will learn
- how the sun supports life on Earth.
- how sunspots and solar flares affect Earth.

What would we do without the sun? Well, we wouldn't need sunglasses, window shades, or sunscreens. These things wouldn't exist because Earth would not exist without the sun.

Even though the sun is 150 million kilometers away, the energy produced by the sun through fusion provides Earth with its heat, light, and other kinds of energy. It provides the energy for photosynthesis, the process plants use to produce food. It provides solar energy, and helps to make fuels like oil and coal. Sunlight also controls Earth's weather. It supplies the heat that causes wind and evaporates water to form clouds and rain.

All life on Earth depends on the constant energy of the sun. Cloudy days occur because of Earth's atmosphere, not because of changes in the sun. If the sun varied a lot, life on Earth would be threatened.

The sun controls Earth's weather.

Solar flares

Astronomers know that the temperature of certain places on the sun does vary. For some time, people have been able to see dark spots on the surface of the sun. These dark spots are called sunspots. **Sunspots** are cool spots on the sun's surface. Sunspots move across the sun. They get bigger and smaller, and at times there are more of them. Some last only a few hours, and others for weeks or months. Two Earths could fit in the space of the average sunspot.

How long do sunspots last?

Near sunspots, giant bursts of fiery gases, called **solar flares**, shoot tens of thousands of miles above the solar surface. Solar flares last from about ten minutes to an hour and are much hotter than the sun's surface.

Scientists are just beginning to find out the effect sunspots and solar flares may have on Earth. Patterns of dry weather seem to be the same as the sunspot cycle. Other people have suggested that cold periods on Earth are related to times when there are few sunspots and flares.

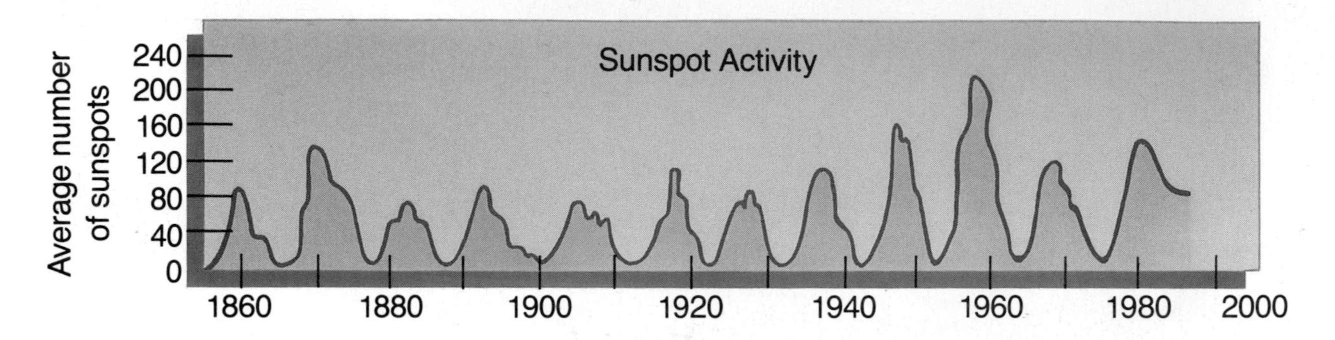

We do know that solar flares affect Earth. Huge solar flares have affected our man-made satellites in space and have been responsible for static in radio lines, some communication blackouts, and power failures. Some people have reported automatic garage doors opening and closing by themselves during these solar flares.

Sunspots are cooler areas of the sun.

How does the solar cycle affect our weather? What controls the cycle? Is the sun changing? These questions are still mysteries to us.

SCIENCE AND . . .
Math

Some weather cycles on Earth are 11 years long. Some are 22 years long. What number sentence is in the same family of facts as $11 \times 2 = 22$?

A. $11 + 2 = 13$
B. $22 \div 11 = 2$
C. $11 + 11 = 22$
D. $22 - 2 = 20$

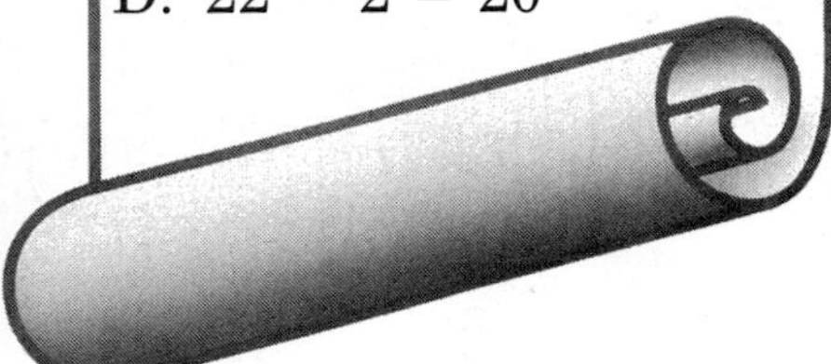

Lesson Summary

- The sun supports life on Earth in many ways, from providing energy for photosynthesis to controlling weather.
- Patterns of dry weather on Earth seem to relate to the sunspot cycle. Cold periods on Earth are sometimes related to periods of few sunspots and solar flares. Solar flares can cause static in radio lines, communication blackouts, and power failures.

Lesson Review

1. What is a sunspot?
2. What are solar flares?
★3. How does the sun control Earth's weather?

Solar Eclipse

LESSON 3 GOALS
You will learn
- what a solar eclipse is.
- what you can see during a solar eclipse.

What would you think if, during a perfectly sunny day, something slowly crossed in front of the sun and blocked its light? Many ancient people must have thought the world was coming to an end. The ancient Chinese used firecrackers and gongs to scare away the dragon they thought was trying to swallow the sun.

A View of a Solar Eclipse

The activity just described occurred during a **solar eclipse** (ih KLIHPS), an event in which the moon comes directly between Earth and the sun. How is it that the smaller moon can block the light of the giant sun? Think about how a small person can block your view of a huge movie screen. The person up close appears larger than the distant screen. Something like this happens when the moon appears to cover the sun during an eclipse. Because the moon is closer to us than the sun, it can block out the sun.

An eclipse can be total or partial. During a total eclipse, the moon completely blocks out the sun for a period of two to seven minutes. Sometimes people can see the other stars during the day when a total solar eclipse occurs. A partial eclipse happens when the moon covers only part of the sun. It is much less noticeable than the total eclipse because the sky does not darken very much.

Because we know the patterns of the rotation of Earth, the sun, and the moon, we can predict when a solar eclipse will occur. But a solar eclipse can't be seen everywhere on Earth at the same time. You know that when it is day where you are, it is night on the other side of Earth. So of course, only people experiencing day can view a solar eclipse. Also, only people in a certain area in the moon's shadow will be able to see a solar eclipse. You have to be in the right place at the right time!

Astronomers are very interested in watching solar eclipses. With special telescopes, they can observe the sun under conditions not normally seen. Parts of the sun can be seen only during a solar eclipse.

A solar eclipse

Why isn't the sun's corona visible every day?

During a total solar eclipse, a bright crown of light flashes into view around the sun. This crown is the sun's **corona** (kuh ROH nuh), the outermost layer of gases surrounding the sun. The word corona is actually related to the word crown. Because the sun's corona is so much dimmer than the rest of the sun, the light of the sun outshines it except during a total solar eclipse. It is from the corona that solar flares shoot out into space.

Table 1 Total Solar Eclipses

Date	Location
7/11/91	Hawaii, Brazil, Central America
6/30/92	South Atlantic Ocean
11/3/94	South America, South Atlantic Ocean
10/24/95	Southern Asia

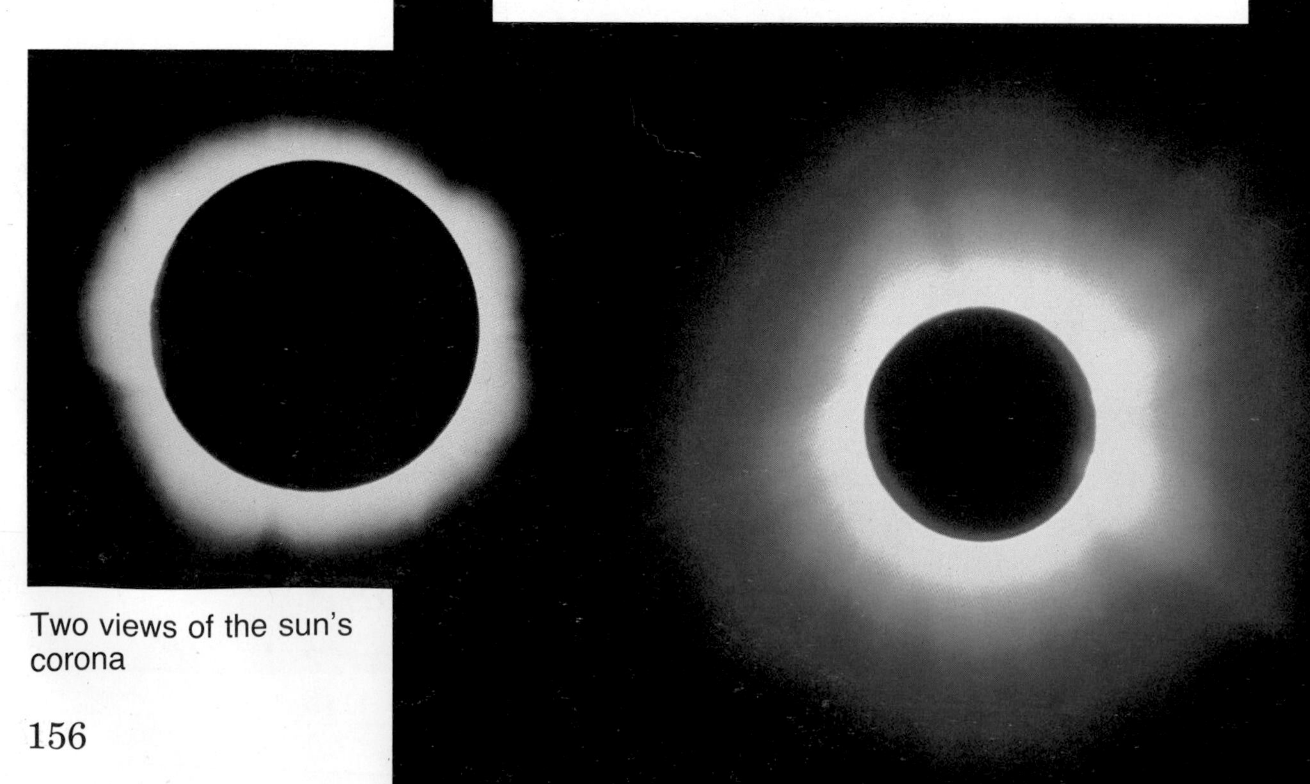

Two views of the sun's corona

Scientists have learned much in their study of the sun and other stars. There are many questions left unanswered. New questions are formed when additional information about the stars is gathered. Perhaps you will be the one to help discover answers to the mysteries of space.

A solar telescope

Lesson Summary

- A solar eclipse is an event during which the moon comes directly between Earth and the sun.
- During a solar eclipse, scientists can study parts of the sun they can't normally see.

Lesson Review

1. How can the moon, which is smaller than the sun, block out the sun's light?
2. What is the sun's corona?

★3. Why can't everyone see an eclipse at the same time?

What causes an eclipse?

What you need

flashlight
ball
globe
pencil and paper

What to do

1. Work with a partner in a darkened room.
2. Stand 2 meters from the globe. Shine the flashlight on the globe. The flashlight represents the sun.
3. Have your partner put the ball between the flashlight and the globe as shown.
4. Record your observations.

What did you learn?

1. What do the ball and the globe represent?
2. Find the center of the shadow on the globe. What would you see if you were standing there?
3. What does this model represent?

Using what you learned

1. How would you arrange the globe, light, and ball to make an eclipse of the moon? Try this lunar eclipse and observe your results.
2. Draw a picture showing:
   a. a solar eclipse.
   b. a lunar eclipse.

I WANT TO KNOW ABOUT...

A Space Engineer

You may have noticed satellite dishes in some people's yards. These dishes can pick up television and radio signals that are sent from a satellite. Most weather reporters today get their information and pictures from satellites orbiting high above Earth.

Edward Harris is an engineer who designs satellites. The weather satellite he is working on will send back information about Earth's oceans. From space, measurements will be taken of water temperature, wind direction and speed, wave height, tides, and ocean currents. The satellite will also carry a camera that will take many small pictures. On Earth, the pictures will be put together into one large picture.

In the last 30 years, satellites have been responsible for great improvements in our communications and weather monitoring. New and improved satellite designs will take us into the future.

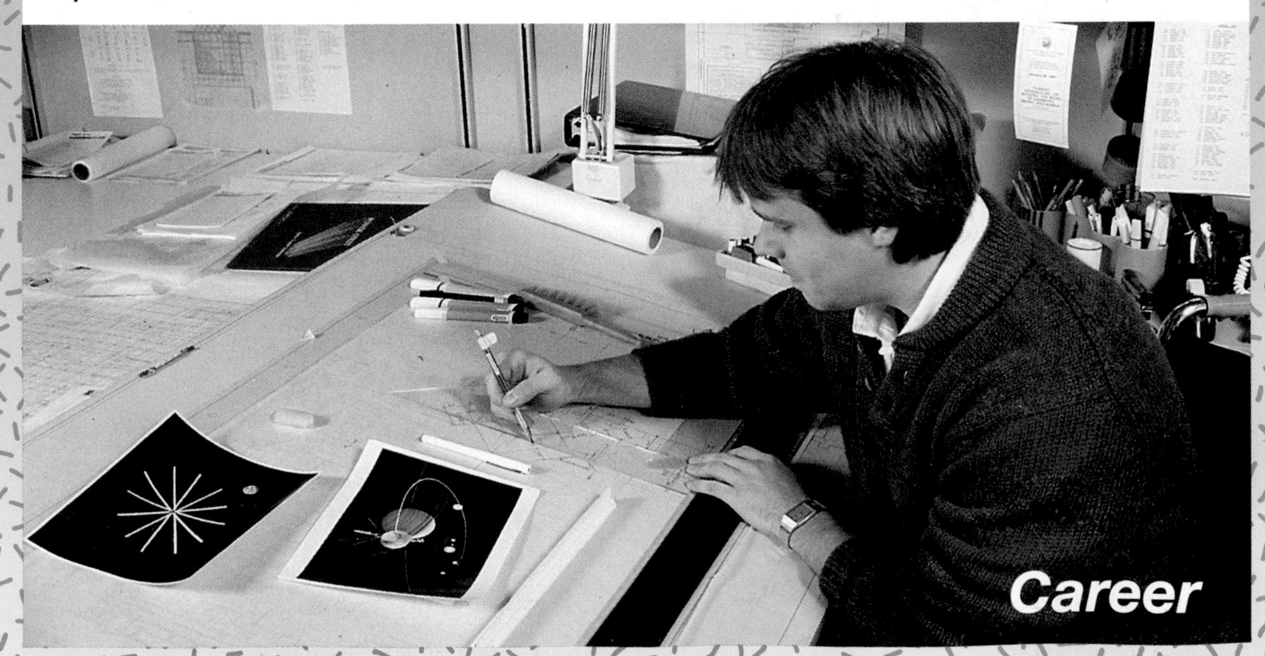

CHAPTER REVIEW 7

Summary

Lesson 1

- The sun is an average-sized yellow star that is five billion years old.
- The sun's core produces energy when hydrogen atoms fuse to make helium.

Lesson 2

- The sun supports life on Earth in many ways, from providing energy for photosynthesis to controlling the weather.
- Patterns of dry weather on Earth seem to relate to the sunspot cycle. Cold periods on Earth are sometimes related to periods of few sunspots and solar flares. Solar flares can cause static in radio lines, communication blackouts, and power failures.

Lesson 3

- During solar eclipse, the moon comes directly between Earth and the sun.
- During a solar eclipse, scientists study parts of the sun they can't normally see.

Science Words

Fill in the blank with the correct word or words from the list.

sun
neutron stars
black holes
red giant
supernova
solar flares
solar eclipse
corona
fusion
sunspots

1. Giant bursts of fiery gases that shoot tens of thousands of miles above the solar surface are ____.
2. Cool spots on the sun's surface are ____.
3. The process of joining atoms together is ____.
4. The moon comes directly between Earth and the sun during a(n) ____ .
5. The layer of gases that surrounds the sun is its ____.

Questions

Recalling Ideas

Correctly complete each of the following sentences.

1. Our sun is ___ star.
   (a) a red (c) a white
   (b) a yellow (d) an orange
2. Scientists believe our sun is about ___ years old.
   (a) one million (c) five billion
   (b) one billion (d) ten billion
3. The last possible stage in a star's life cycle is a
   (a) supernova. (c) neutron star.
   (b) black hole. (d) white dwarf.
4. The moon only covers part of the sun in a
   (a) partial eclipse.
   (b) total eclipse.
   (c) solar flare.
   (d) sunspot.

5. The moon completely covers the sun in a
   (a) partial eclipse.
   (b) total eclipse.
   (c) solar flare.
   (d) sunspot.

Understanding Ideas

Answer the following questions using complete sentences.

1. Why do some stars in the night sky appear brighter than others?
2. How does the sun produce energy?
3. How does the sun support life on Earth?
4. Describe ways in which solar flares affect Earth.
5. Why is it that the sun's corona can best be studied during a solar eclipse?

Thinking Critically

Think about what you have learned in this chapter. Answer the following questions using complete sentences.

1. Will our sun become a supernova? Explain your answer.
2. How would life on Earth be affected if the sun's energy was not constant as it is now, but varied greatly instead?

Oceans and Seas

LESSON 1 GOALS
You will learn
- what the major oceans of the world are.
- how the oceans affect our climate.

In 1519, Ferdinand Magellan led the first sailing expedition that made it completely around the world. Magellan's voyage changed scientific thinking forever. He proved that Earth was round and larger than anyone had imagined. He also proved that the oceans of the world are connected.

Oceans of the World

What is the largest ocean on Earth?

What Magellan helped us learn was that 70 percent of Earth's surface is made of connected bodies of salt water, called **oceans.** The largest of these, the Pacific Ocean, covers about one third of Earth's surface. The Atlantic is about half the size of the Pacific. The Indian Ocean is smaller than the Atlantic. These three oceans come together around Antarctica, at the southern-most part of Earth. The Pacific and the Atlantic come together near the North Pole, forming the Arctic Ocean.

Oceans of the world

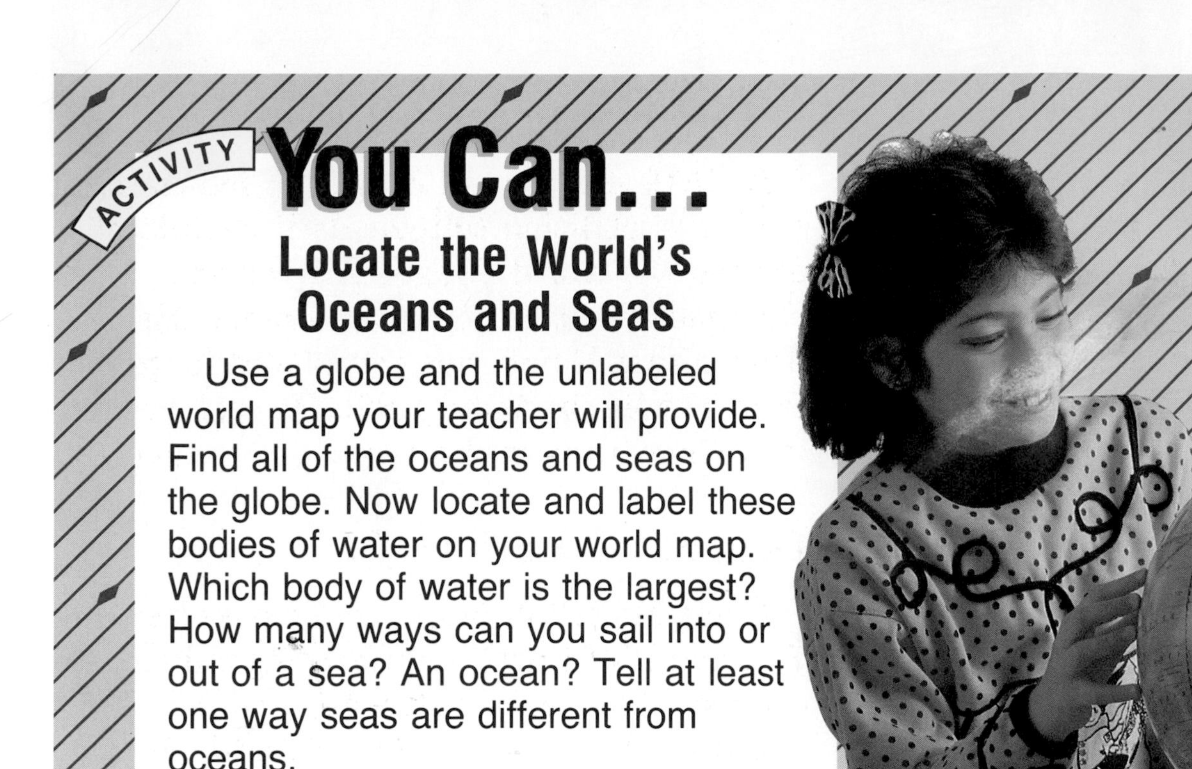

ACTIVITY

You Can...

Locate the World's Oceans and Seas

Use a globe and the unlabeled world map your teacher will provide. Find all of the oceans and seas on the globe. Now locate and label these bodies of water on your world map. Which body of water is the largest? How many ways can you sail into or out of a sea? An ocean? Tell at least one way seas are different from oceans.

Some smaller areas of water in oceans are called **seas.** For example, the Caribbean Sea, between North America and South America, is a part of the Atlantic Ocean. Some seas, like the Mediterranean Sea, are surrounded by land with only one opening to an ocean. Sailing across the oceans and seas, people have found many other water routes around the world.

Ocean breezes affect land temperatures.

Oceans and Land Climates

How do climates near the ocean compare to inland climates?

The oceans and seas of the world have a large effect on our weather. Climates near the ocean usually have less extreme temperature differences between summer and winter than inland climates. This is because water heats and cools more slowly than land does.

On hot summer days, land heats quickly while the oceans remain cool. The ocean winds blow cool breezes over the shore in the afternoon. These ocean winds keep the temperatures lower in the summer on land near oceans. Early in the morning, the land may cool down so much that it is actually cooler than the ocean. When this happens, the cool night air from the land blows out over the ocean where it is warmed up. In this way, the ocean is a stable influence on land temperatures.

The temperature of ocean water does vary around the world, but it's constant enough to control land temperatures. The surface ocean water near the equator gets as warm as 28°C. The surface water at the North and South Poles may be as cold as −1°C. At both poles, the deeper water is about the same temperature as water at the surface. But in most of the ocean, deeper water is colder than water at the surface. The water at the bottom of the oceans of the world has an average temperature of about 3 to 4°C.

Would You Believe?

If all the water in the oceans was shared equally among every person in the world, each person would have enough water to fill 80 swimming pools.

Temperature of ocean surface water varies.

Oceans also affect the climate by providing Earth with rain and snow. When the sun heats the ocean, some of the surface water evaporates. It rises and forms clouds. The winds carry the clouds across the land. Rain or snow falls from the clouds, bringing water to plants and filling rivers, lakes, streams, and ponds.

What is one important event that occurred in 1986?

We have come a long way since Magellan's time. Instead of taking four years, we can now fly around the world in much less time. In 1986 Dick Rutan and Jeana Yeager made the first nonstop flight around the world. The flight took just over nine days (216 hours, 3 minutes, 44 seconds). Even with these advances, we still have a lot to learn.

Ocean water evaporates and returns to Earth as rain or snow.

The oceans control our life on Earth in ways we are only beginning to understand. Perhaps you will be the one to unlock more of the mysteries of the deep.

Advances have been made in world travel.

Lesson Summary

- Earth's three major oceans are the Pacific, the Atlantic, and the Indian Oceans.
- Oceans affect Earth's climate by stabilizing temperatures and providing moisture.

Lesson Review

1. What are oceans?
2. How do oceans control the temperature on land?

★3. What would Earth be like if there were no oceans?

Ocean Movements

LESSON 2 GOALS
You will learn
- what ocean waves, tides, and currents are.
- how waves, tides, and currents are formed.

If you have ever been to an ocean shore, you have heard the rhythmic sound of the waves hitting the beach. Ocean water is always in motion. It is this rhythmic sound of waves rushing over rocks that attracts people to the beach.

Waves

Ocean waves are the rising and falling of water. They can be caused by wind, earthquakes, or the pull of the moon's and sun's gravity.

What are the crest and trough of a wave?

You may have heard about surfers riding the crest of a wave. Waves have a high point and a low point. The **crest** of a wave is its high point. The **trough** is its low point.

Ocean waves slow down and begin to drag as they move into shallow water close to shore. As they slow down, the crests get closer together. This causes the waves to get steeper until they fall over. We say the wave "breaks" onto the shore.

In search of the perfect wave

Tsunamis can reach a height of 30 meters.

Earthquakes on the ocean floor can set huge waves in motion. These are called tidal waves or **tsunamis** (soo NAHM eez). Most tsunamis start in Japan, Alaska, or Chile, where earthquakes are frequent. Tsunamis can travel as fast as 700 kilometers per hour. In the ocean, they are difficult to see, but in shallow water the waves become dangerously high.

Tides

Why is it that castles you build in the sand at the beach are always gone the next day? If you had the time to watch, you would notice that the water rises and then goes back down over the course of about 12 hours. Sometimes the water completely covers the beach. The rise and fall of ocean water levels is called a **tide**.

Tides are caused by the pull of the moon's gravity. When the moon is directly over water, it pulls the water toward it. This creates a bulge of water on Earth. On the opposite side, the Earth is pulled toward the moon more than the water. Therefore, two "bulges" form in the oceans.

SCIENCE AND . . .
Writing

Rewrite the following correctly. Waves transmit energy. Through ocean water.

A. Waves transmit. Energy through ocean water.
B. Energy waves transmit through ocean water.
C. Waves transmit energy through ocean water.

Two places on Earth move through these bulges as Earth rotates. This causes the water level to rise at these places. This is called high tide. Low tides occur when each place moves out of the bulge and the water level goes down. The tidal bulge is like the crest of a very large wave that travels around the world as Earth rotates.

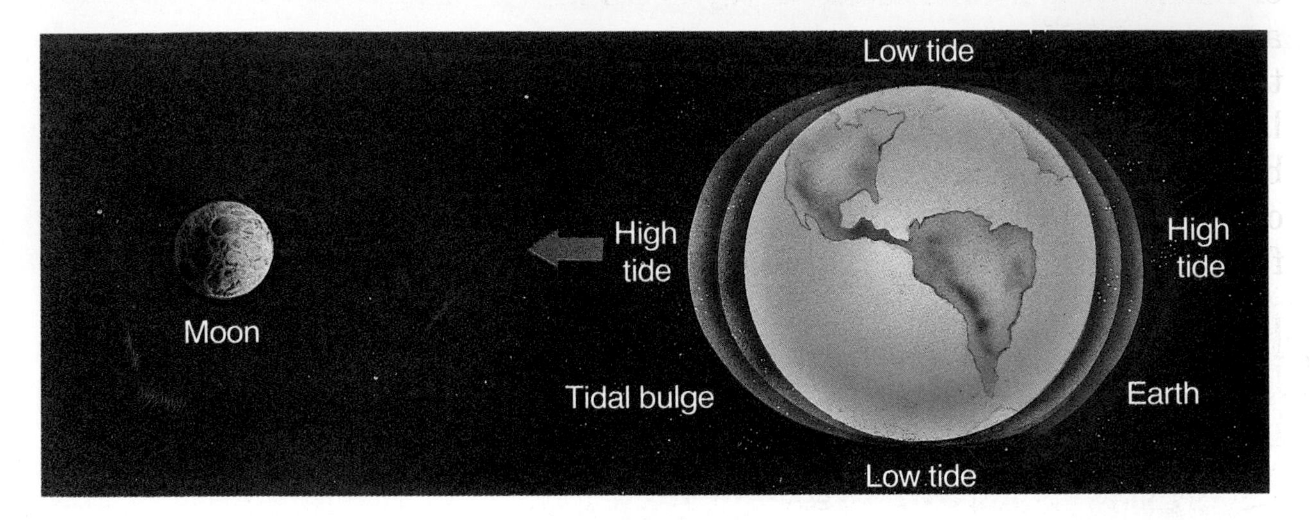

Tidal bulges

The shape and size of the ocean basin also affect the tides. Most places on Earth have two high tides and two low tides each day. But, because of the shape of the ocean basin, some places have only one high tide and one low tide each day.

Tides are also affected by the position of the sun, moon, and Earth. When the gravitational pull of the sun and moon are in the same direction, the difference between high and low tide is great. When the sun and moon do not pull in the same direction, the difference between high and low tide is small. Although the sun's pull of gravity on Earth affects the ocean tides, the sun is much farther away from Earth and has less effect on tides than the moon.

Why does the sun have less effect on tides than the moon?

Currents

Ocean water moves from place to place around the world. Some scientists think that one part of water could move through all the oceans over the course of 5,000 years. Waves and tides move water vertically, or up and down. The side to side, or horizontal, flow of ocean water is called a **current**. Sometimes you can walk through a cold current while wading at a lake or at the beach. You may have also been warned about dangerous undertows, or underwater currents, that can pull you far out in the water.

What is an undertow?

Dangerous undertow

Wind blowing across the water's surface can cause a current. General wind patterns in the atmosphere produce large ocean currents that are in constant motion. Certain surface currents move warm water from near the equator toward the north and south poles. The Gulf Stream is one of these currents. It carries warm water from the Caribbean Sea northward along the eastern coast of the United States.

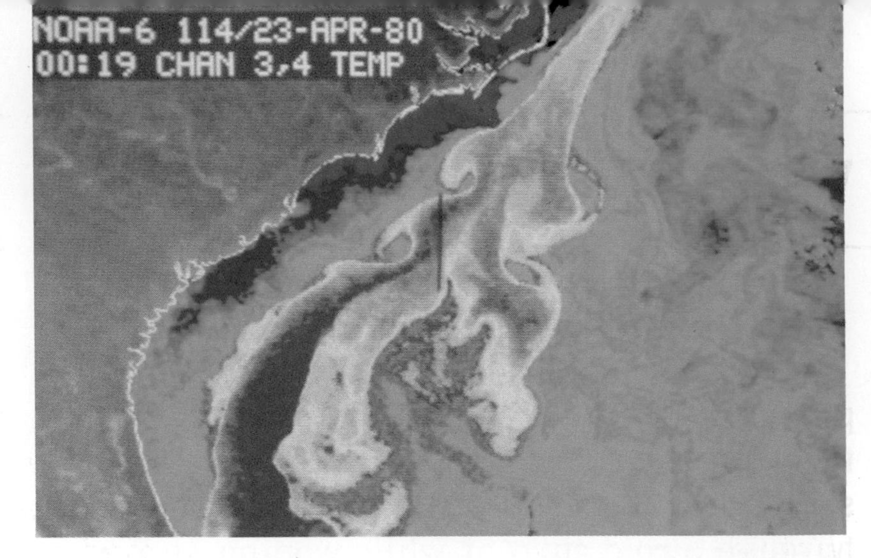

Gulf stream currents

Other surface currents flow toward the equator. They carry cold water to the equator from areas nearer the poles. There are also deep water currents. These currents help balance the flow of water caused by the surface currents. Deep water currents move cold water from the poles toward the equator.

Lesson Summary

- Water is always in motion. The rising and falling of water caused by wind are waves. The rise and fall of ocean water levels caused by the moon's gravitational pull are called tides. The horizontal movements of water are currents.
- Ocean water movements are caused by wind, gravity, earthquakes, volcanoes, or other movements on the ocean floor.

Lesson Review

1. What is a tsunami?
2. What causes tides to rise and fall?
★**3.** How do currents help to keep land temperatures stable?

What causes deep ocean currents?

What you need

plastic storage box
tap water
small plastic bag
twist tie
food coloring
hot tap water
dropper
rock
ice cube
pencil and paper

What to do

1. Fill the box 3/4 full of tap water.
2. Place the rock in the plastic bag. Fill the bag 1/2 full of hot water. Tie the bag closed.
3. Place the bag on the bottom of one corner of the box.
4. Float the ice cube in the opposite corner.
5. Add 4 drops of food coloring to the water next to the ice cube.
6. Observe the food coloring for 5 minutes. Record your observations in a table.
7. Draw what you observe.

What did you learn?

1. Where was the water colder? Warmer?
2. Describe the movement of the food coloring over several minutes.

Using what you learned

1. How is your container of water like the ocean?
2. The oceans are coldest at the poles and warmest at the equator. How might ocean water move between the poles and the equator?
3. Why does cold water sink?
4. Why does warm water rise?

Ocean Resources

LESSON 3 GOALS
You will learn
- what ocean water is made of.
- what natural resources can be found in the ocean.

"Water, water, everywhere,
Nor any drop to drink."

Samuel Taylor Coleridge wrote these famous lines in his poem "The Rime of the Ancient Mariner." He was describing what it was like to be out in the middle of the ocean. Before airplanes, sailing was the only way people could get around the world. They knew that although the ocean was filled with water, this water would not quench their thirst. What is it about ocean water that makes it like this?

Ocean Water

If you taste ocean water, the first thing you will notice is its saltiness. Although there are many kinds of material dissolved in ocean water, the most abundant material is the same as plain table salt. Around the world, ocean water is about three and one-half percent salt. The movement of the waves and currents keeps the water pretty much the same.

If you eat a lot of salty chips, you get thirsty. You may even wake up in the middle of the night dying of thirst. This is because your body uses water to get rid of salt. The more salt you take in, the more water your body needs to get rid of it. Drinking ocean water only increases thirst because it has so much salt in it. If you only drank ocean water, you would lose too much water and become sick.

How did ocean water become salty over a long period of time?

How did salt get into ocean water? You know that Earth's water is used over and over in the water cycle. It moves from place to place by evaporation, condensation, and precipitation. Rain and snow that fall to Earth run back into the ocean and carry with them a tiny bit of dissolved rock. This dissolved rock has salt and other minerals in it. The amount of dissolved material carried to the ocean is very small. Over a long time, however, it collects in large amounts. Because only pure water evaporates, when water evaporates from the ocean, the salt stays behind and builds up over time.

Salt from ocean water

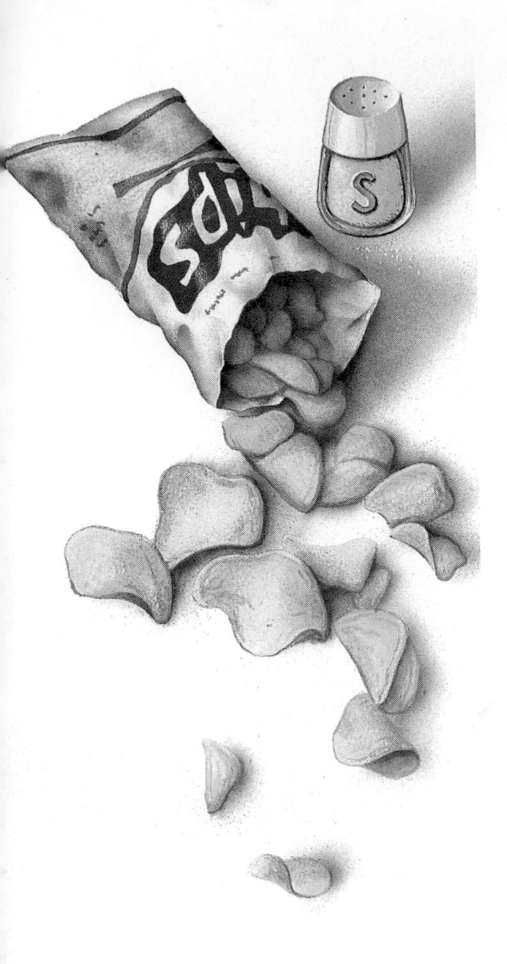

The salt you use at home was probably once dissolved in ocean water. Salt is collected for people's use in a variety of ways. In some places, ocean water is poured into large ponds on land. The water evaporates leaving the salts behind.

Removing salt from ocean water to make it drinkable is a more complex and expensive process. Ocean water is evaporated and then condensed back to liquid water.

Production of fresh water from ocean water

Salt is not the only mineral in ocean water. All the elements that make up the minerals on Earth are in ocean water.

Natural Resources

What is a natural resource?

A **natural resource** is a material that nature supplies that is valuable to us. Think of some common natural resources. Trees, coal, oil, and water might be on your list. People are really just beginning to understand how to harvest the natural resources that the ocean provides.

The first things you might think of when you consider ocean harvests are animals and plants. Around the world seafood is a major part of people's diets. You can probably name several animals that live in the sea and are commonly eaten by people. Tuna, lobster, crab, and shrimp are just a few. In many parts of the world, people also eat plants from the sea.

Ocean plants and animals can provide natural resources other than food. For example, **kelp** is a large, tough, brown seaweed. The ashes of kelp are a source of iodine. You may have used iodine on a scrape or cut in your skin. Kelp is also used in making dyes and in photography.

Some ocean animals remove certain minerals from ocean water. These minerals make up the skeletons and shells of the animals. When the animals die, their skeletons and shells remain. Over many years, thick layers of these skeletons build up. **Limestone** is a rock formed from these skeletons, and is commonly used as a building material.

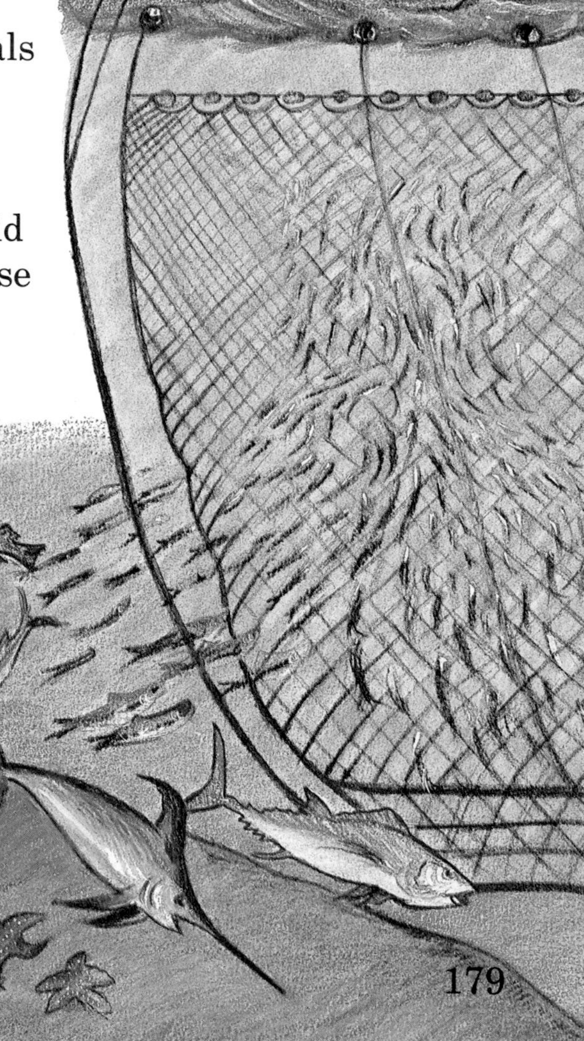

Some minerals are found on the ocean floor in small, round lumps. These lumps are called **nodules** (NAHJ ewlz). Nodules can be scooped up from the ocean floor. Manganese (MANG guh neez) and nickel are two valuable metals found in nodules.

What two valuable metals are found in nodules?

Manganese is often used in the steel industry. It removes impurities from steel to make a clean metal. Manganese is also used in some paints, dyes, and fertilizers. Nickel mixed with other metals builds a stronger metal that lasts longer. The U.S. five-cent coin is usually made of 25 percent nickel and 75 percent copper.

You know that people get oil and natural gas by drilling wells on land. We can also get oil and natural gas from the ocean. From large platforms, oil companies drill wells into the ocean floor. Ships or pipelines then transport the oil and natural gas to shore.

Deep currents rising toward the surface also bring nutrients up into the surface waters. The nutrients are used as food by living things in the ocean.

Offshore drilling for oil and natural gas

Besides minerals, oil products, and nutrients, there are many important gases in ocean water. Two of the most important gases are oxygen and carbon dioxide. They are important to life in the ocean as well as life on land.

Water—can't live without it

What two important gases are found in ocean water?

The oceans and seas of the world supply us with many natural resources. To make sure these resources survive, we must protect these bodies of water.

Lesson Summary

- Ocean water contains dissolved minerals and salts.
- Oceans provide natural resources, including food, minerals, and fuel.

Lesson Review

1. How do minerals get into ocean water?
2. Name four natural resources that can be taken from the ocean.

★3. Why might it be difficult to find and get natural resources from the ocean?

Why is it easier to swim in the ocean?

What you need

tall clear narrow jar
pencil with eraser
metric ruler
"fresh water"
thumbtack
ballpoint pen
paper towel
"ocean water"
pencil and paper

What to do

1. Fill the jar to within 1 cm of the top with "fresh water."
2. Put a thumbtack into the pencil eraser, as shown. Put the pencil, eraser down, into the water. Use the pen to mark the water level on the pencil.
3. Remove the pencil and dry it. Measure the length, in millimeters, of the pencil above the water level mark.
4. Repeat steps 1 to 3 using the "ocean water."

What did you learn?

1. In which water did the pencil float higher?
2. Predict how the pencil will float if you add a spoonful of salt to the "ocean water." Try it and record your results.

Using what you learned

1. Why is it easier to swim in the ocean than in fresh water?
2. Can a boat carry a heavier load in fresh water or salt water? Why?

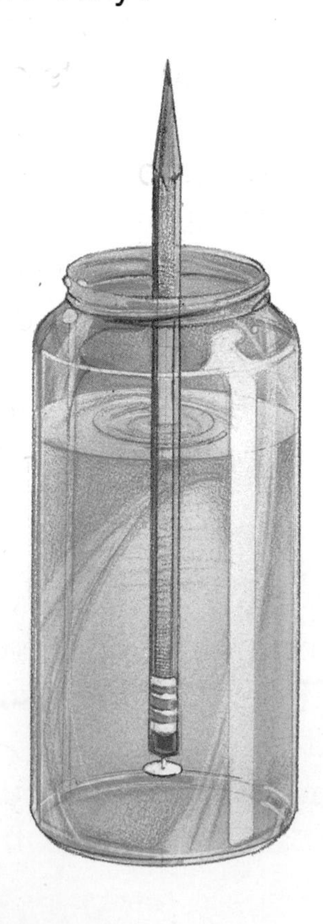

I WANT TO KNOW ABOUT...

Diatoms

You don't have to be big to be powerful. You can't even see the most important organism in the ocean, but it is everywhere.

Diatoms (DI uh tahmz), tiny one-celled golden algae, live in the ocean and some fresh waters. There are hundreds of thousands of diatoms in a bucket of ocean water.

Diatoms are important organisms in the ocean. They trap the sun's energy and produce large amounts of oxygen and food for other sea life. A large humpback whale may feed on several hundred billion diatoms every few hours.

When diatoms die, their tiny silica shells drop to the ocean floor in layers called ooze. Millions of years of ooze build-up has made the oceans a good source for silica, which is often used in building materials.

Some of these layers of ooze are also found on land where oceans once flowed. These deposits are called diatomite mines.

Ground diatomite powder is used to polish and clean metals. It is also used in sound and heat insulation and to make dynamite. Another use for diatomite is to filter liquids such as fruit juices, liquid soap, and vegetable oils.

Diatoms are one of many microscopic organisms that have a major effect on life on Earth. Future technology may allow us to understand more of the world beyond our sight.

CHAPTER REVIEW 8

Summary

Lesson 1

- Earth's major oceans are the Pacific, Atlantic, and Indian Oceans.
- Oceans affect Earth's climates by stabilizing temperatures and providing moisture.

Lesson 2

- Water is always in motion. The rising and falling of water caused by wind are waves. The rise and fall of ocean water levels caused by the moon's gravitational pull is called tides. The horizontal movements of water are currents.
- Ocean water movements are caused by wind, gravity, earthquakes, volcanoes, or other movements on the ocean floor.

Lesson 3

- Ocean water contains dissolved minerals and salts.
- Oceans provide us with natural resources including food, minerals, and fuel.

Science Words

Fill in the blank with the correct word or words from the list.

oceans	**tide**	**ocean waves**	**kelp**
seas	**limestone**	**crest**	**nodules**
tsunamis	**natural resource**	**trough**	**current**

1. The ___ of a wave is its high point.
2. The rise and fall of ocean water levels is called a ___.
3. The horizontal flow of ocean water is called a(n) ___.
4. ___ are the rising and falling of ocean water.
5. The ___ of a wave is its low point.

Questions

Recalling Ideas

Correctly complete each of the following sentences.

1. Large connected bodies of salt water are called
   (a) lakes. (c) rivers.
   (b) seas. (d) oceans.
2. The amount of Earth's surface covered by oceans is
   (a) 30 percent. (c) 70 percent.
   (b) 50 percent. (d) 95 percent.
3. Waves caused by earthquakes are
   (a) currents. (c) tides.
   (b) tsunamis. (d) all of these
4. Tides are mainly caused by the pull of
   (a) Earth's gravity.
   (b) the moon's gravity.
   (c) the sun's gravity.
   (d) the other planets.
5. Ocean water moves from side to side in
   (a) waves. (c) tides.
   (b) currents. (d) all of these

Understanding Ideas

Answer the following questions using complete sentences.

1. Name and compare the three major oceans on Earth.
2. How do oceans affect climate?
3. What causes ocean waves?
4. Explain the changes that occur on Earth's surface during high and low tides.
5. What causes ocean currents?

Thinking Critically

Think about what you have learned in this chapter. Answer the following questions using complete sentences.

1. What would Earth be like if oceans only covered about 20 percent of its surface?
2. Why is surfing best at the parts of the ocean that have level, gently-sloping beaches?

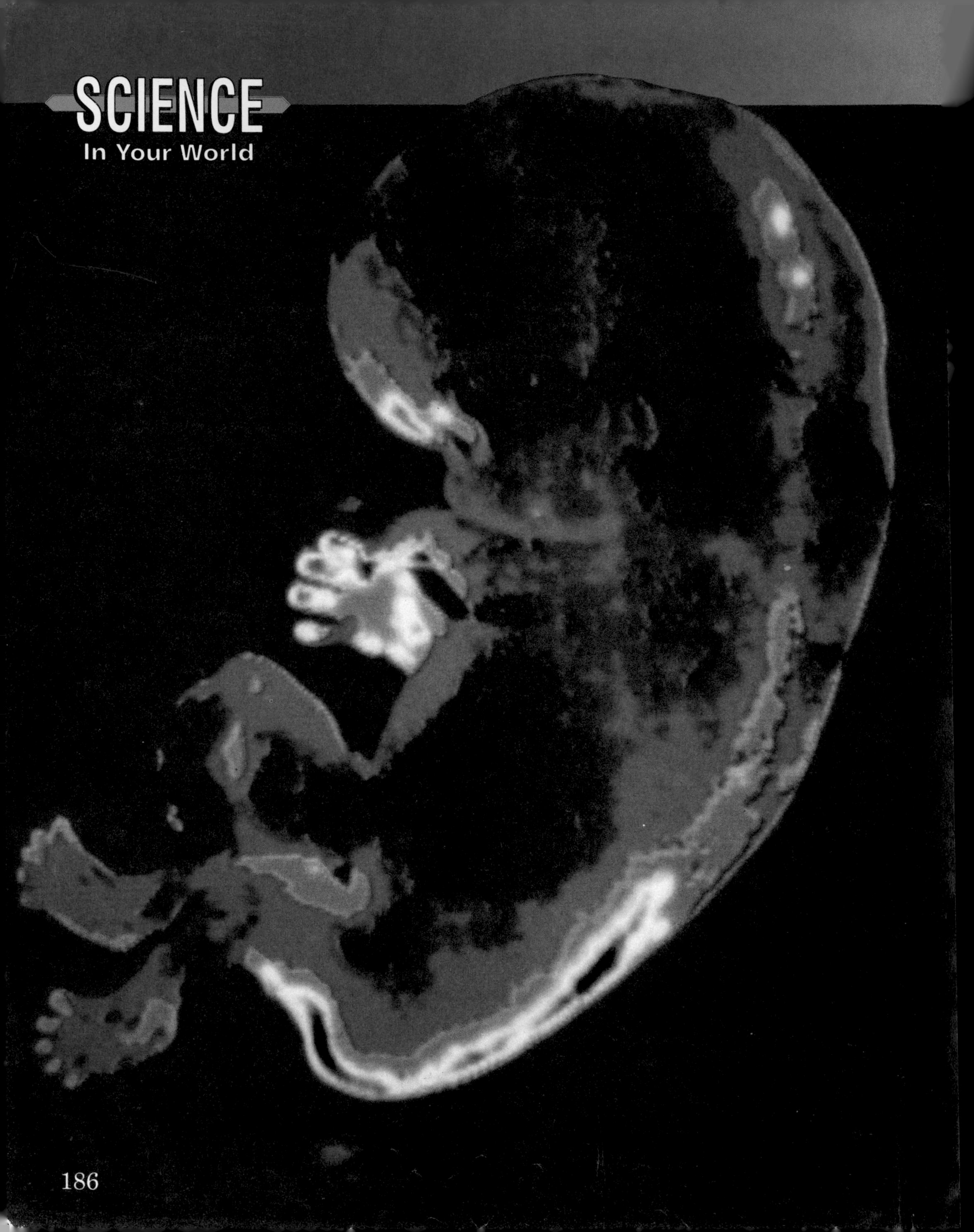
SCIENCE
In Your World

Exploring the Ocean

Your mother's doctor may have used ultrasound to get a picture of you before you were born. In ultrasound, sound waves travel through fluid in the mother's body. The sound waves reflect back. These echoes are used to create a picture of the baby. Scientists use something similar called sonar to map the surface of the ocean floor.

Have You Ever...

Heard an Echo?

Go to a large, empty room. What do you think will happen if you shout hello? Try it. What do you hear? Will you hear an echo in your classroom, an empty hallway, between outside buildings, or near a brick wall? Try it. Why can't you hear an echo in some places?

Ocean Features

LESSON 1 GOALS
You will learn
- why scientists study the ocean floor.
- how scientists study the ocean floor.
- that the ocean floor has specific features.

If you could drain all the water from the ocean, what would you see? Strange as it may seem, you would see some very large mountains, as well as plains, canyons, ridges, and trenches, on the ocean floor. In fact, the ocean floor has many of the same features you would find on land.

What is sonar?

Although it is impossible to drain water from the ocean, scientists have "seen" much of what the ocean floor is like. They use **sonar**, or reflected sound waves to find out how deep the water is and where objects are in the water. The scientists measure the depth of the ocean by how long it takes the sound waves to return to the ship. They record many depth measurements and mark the changes on a graph. The graph becomes a map of the ocean floor because it shows where some places are deeper than others. By using sonar, scientists can locate many features of the ocean floor.

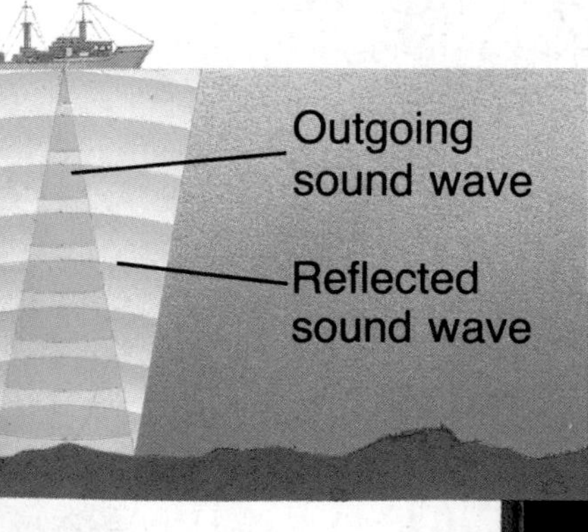

Using sonar to map the ocean floor

The ocean floor has three main parts—continental shelf, continental slope, and plain. You can compare the shape of the ocean floor to a large bowl. The outer edges of the bowl are like the continental shelf. The sides of the bowl are similar to the continental slope. The bottom of the bowl is like the plain. Let's look at each of these features to get a clearer picture of the ocean floor.

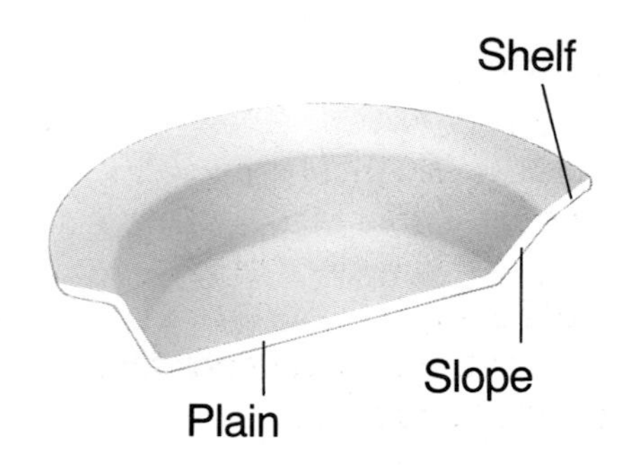

What are three main parts of the ocean floor?

If you have been to the ocean, perhaps you have walked along the beach and into the water. The farther you walk, the deeper into the water you step because of a gentle slope of the land under the water. This gently sloping part of land covered by ocean water is the **continental shelf.** In some places the continental shelf extends hundreds of kilometers, while in other places it is very narrow.

The great Barrier Reef is part of the continental shelf.

Fishing above the continental shelf

The water above the continental shelf is very important to the fishing industry because a lot of fish and organisms such as lobsters and clams live there. In fact, it is the most important area for fishing in the world. The shelf is also a source for oil and natural gas.

What steep-sided features are found within the continental slope?

The **continental slope** begins where the continental shelf suddenly drops off and the water becomes very deep. Many canyons with steep sides are found in the slope. The canyons probably are made as sand, mud, and water flow down the slope to the ocean bottom. Some of the canyons on the continental slope are deeper than the Grand Canyon. Most of the animals that live here feed on animals that have washed down from the continental shelf.

The bottom of the deep, dark ocean is the **plain**. Currents carry mud and sand down the continental slope and deposit this material on the plain. The water on the ocean bottom is so deep, cold, and dark that few organisms can live here.

Forces within Earth form some features on the ocean floor. In some places on the plain, there are undersea volcanoes. Some of these undersea volcanoes grow so large that they rise above sea level and become islands. Did you know that the Hawaiian Islands were formed by undersea volcanoes?

Some of the largest mountain chains are found along cracks in the ocean floor called **rift zones.** Molten material called magma comes up through the cracks and cools to make a new ocean floor. Over time, mountain chains called mid-ocean ridges form along the rift zones. Mid-ocean ridges are found in all the major oceans. The Mid-Atlantic Ridge runs from near Greenland to below the tip of Africa. The Azore Islands in the north Atlantic Ocean are examples of mountain peaks from the Mid-Atlantic Ridge that formed islands.

SCIENCE AND . . . Writing

Which type of mistake, if any, is underlined?
Fish are plentiful along the contenental shelf, but they must be protected.

A. Spelling
B. Capitalization
C. Punctuation
D. No mistake

The ocean floor

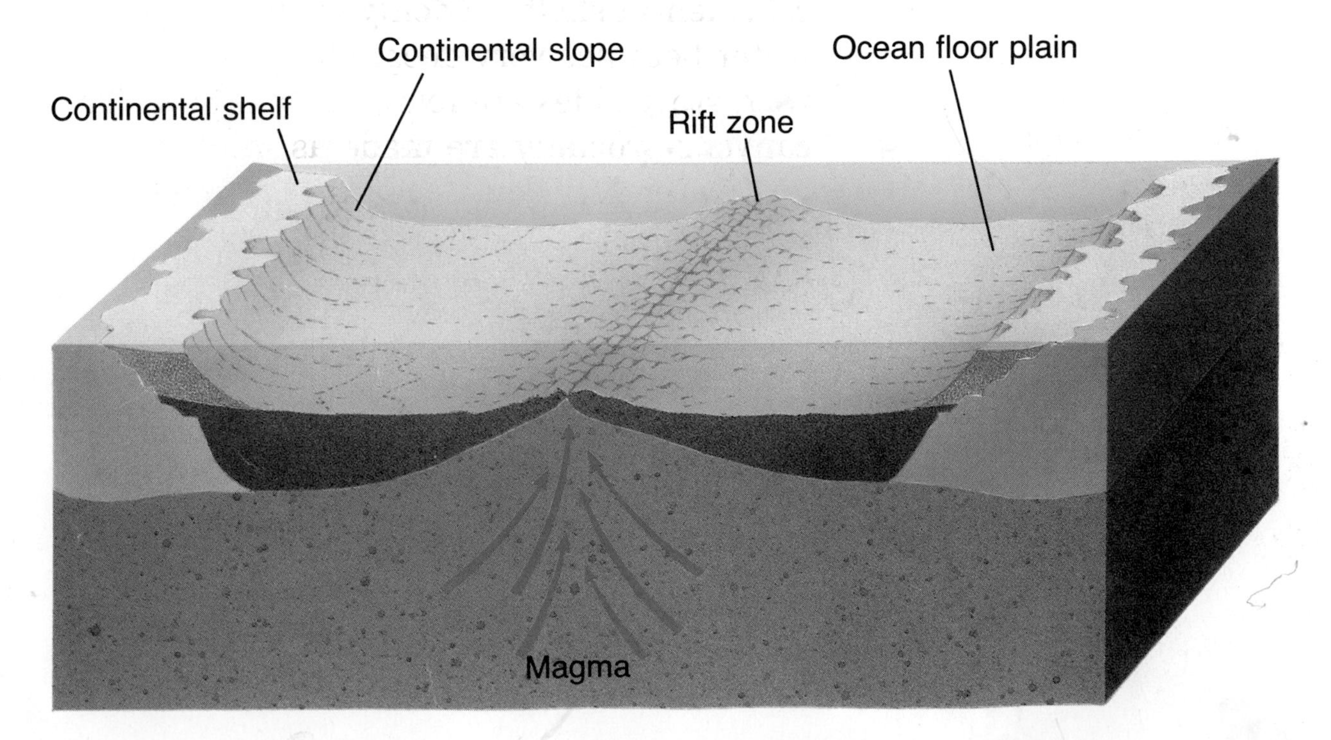

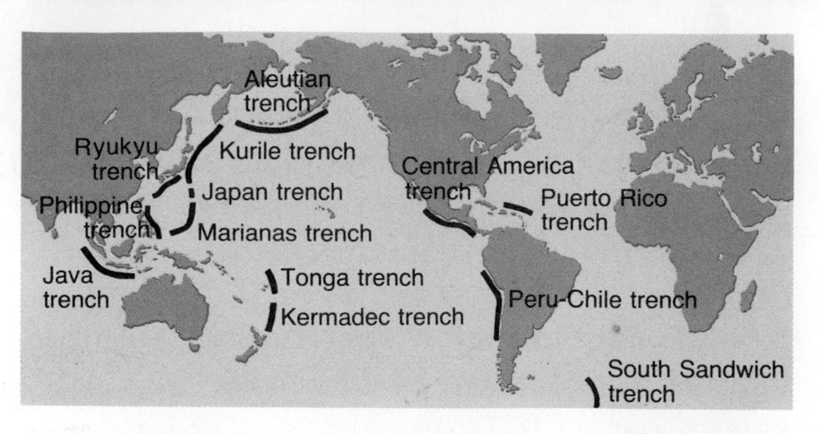

The world's main ocean trenches

What is significant about the Marianas Trench?

The Pacific Ocean has some very deep narrow valleys called **trenches**. Trenches form where the ocean floor is forced down below the continental crust. These trenches are longer and deeper than any canyons found on land. The deepest spot is in the Marianas Trench. It is 11,022 meters below the surface of the Pacific. This is a greater distance than the height of Mt. Everest, the tallest mountain on land.

Scientists have learned many things about the ocean floor. By using the information they have found, we can know how to better use and preserve the resources the oceans provide.

Lesson Summary

- The ocean is a valuable resource.
- Sonar is used to study the ocean floor.
- Features of the ocean floor include the continental shelf, continental slope, plain, rift zones, and trenches.

Lesson Review

1. How does sonar measure ocean depth?
2. What are three parts of the ocean floor?
★3. Name three natural resources that are found within the continental shelf.

How can you map the ocean bottom?

What you need

ocean bottom model box
drinking straw
pencil and paper

What to do

1. Make a table and graph like the ones shown for the row of holes.
2. Carefully probe the straw straight down into each hole of the box until it stops.
3. Measure and record in the table how deep the straw goes into each hole.
4. Record each hole's depth on the graph. Connect the points with a straight line.
5. Compare your drawing to the "ocean floor" in the box.

What did you learn?

1. How was the straw probe used to observe the inside of the box?
2. Describe how your drawing on the graph compared to the "ocean floor" in the box.

Using what you learned

1. What senses did you use to observe the floor of the model box?
2. How is this activity similar to the way the ocean bottom is actually mapped?

Hole Depth Graph

Hole		1	2	3	4	5	6	7	8
Depth (cm)	0								
	2								
	4								
	6								
	8								
	10								
	12								

Hole depth table

Hole	1	2	3	4	5	6	7	8
Depth (cm)								

Ocean Life

LESSON 2 GOALS
You will learn
- that ocean life exists as drifting, free-swimming, or bottom-dwelling organisms.
- that each living thing is adapted to life at a certain part of the ocean.
- that each living thing in the ocean is part of a food cycle.

John and his family look forward to visiting the ocean each year. John especially likes to go fishing. Some days they rent a boat and hire a fishing guide. The guide always seems to know where to go to find lots of different fish. How do you think the fishing guide knows where to find the fish?

Charter boat for fishing

When you studied plants and animals, you learned that living things need certain things to stay alive. Do you know that both plants and animals live in the ocean? Plants are food for many kinds of ocean life, but they can't live in the deep, dark waters of the ocean plain. They need sunlight and nutrients to live. Animals, however, live almost everywhere in the ocean. They depend on plants and other animals to provide the food they need.

We can divide ocean life into three groups—plankton, nekton, and benthos. The members of each group have certain features that make it possible for them to live where they do. Some float on or near the top of the water, while some attach themselves to the ocean bottom. Some forms of ocean life swim freely.

Into what three groups can ocean life be divided?

Organisms that drift and move along with the help of surface currents are **plankton** (PLANG tun). Many of them are so small that they can't be seen without a microscope. Some plankton are like plants and can make their own food. They live close to the ocean's surface so they can get the sunlight to grow. Animallike plankton also are at the surface. They feed on the plantlike plankton.

Diatoms are the most important of the plantlike plankton group. They are yellow-green algae that provide most of the food for ocean life. When storms or swift currents stir up nutrients from deep ocean water, the diatoms grow very quickly. They attract large number of animals that feed on them.

Diatoms

The plankton group also includes animals such as crab larvae and jellyfish. They depend on surface currents to move.

Many of these animals come to the surface of the ocean at night to feed on the plantlike plankton, and then sink lower into the ocean during the day.

What are nekton?

Larger free-swimming organisms like sea turtles and fish are **nekton** (NEK tun). Over 13,000 kinds of nekton swim about freely in the ocean. They range in size from small herring to large whales. Nekton also includes octopuses, squids, and seals.

Although these animals can swim about freely, they must live in certain parts of the ocean. Some prefer cold temperatures, while others prefer warmth. They must also live where they can get the food supply they need and where the amount of salt in the water is just right.

Three groups of ocean life

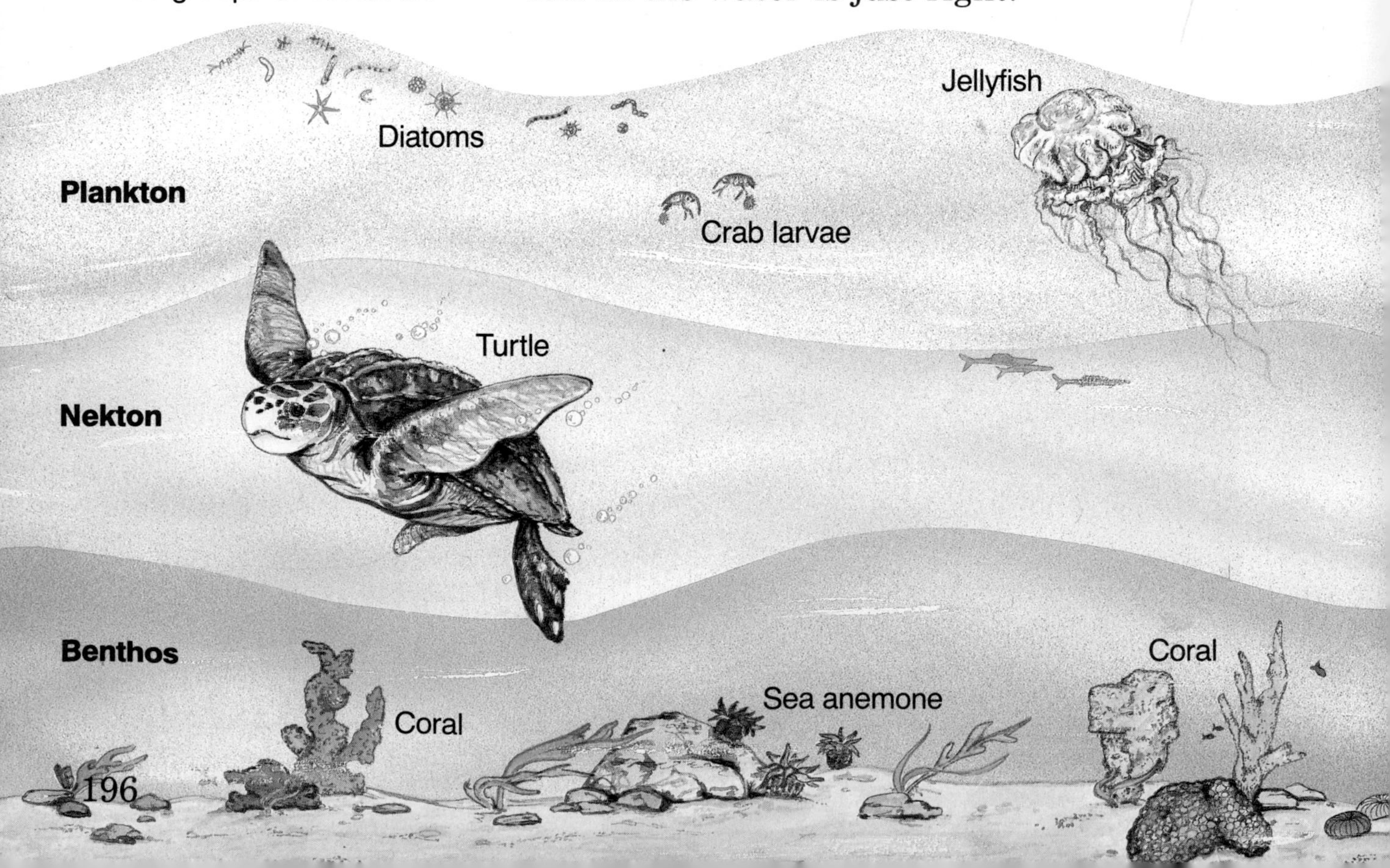

Many of them come to the surface at night to feed on the plankton. Others, like the angler fish, rely on their unusual body parts to attract other fish for food. The angler fish has an antennalike organ on the top of its head. It can make the tip of this organ glow. When a curious animal comes too close, the angler fish catches its dinner.

Benthos (BEN thas) are plants and animals that live on the ocean bottom. Some, like barnacles and oysters, attach themselves to the ocean bottom. Clams and other shelled animals burrow into the mud and sand. Ocean currents carry their food down to them.

Why are barnacles and oysters classified as benthos?

Corals live in warm water. They build structures from the minerals they get from seawater. The structures join together to make a colony, or coral reef.

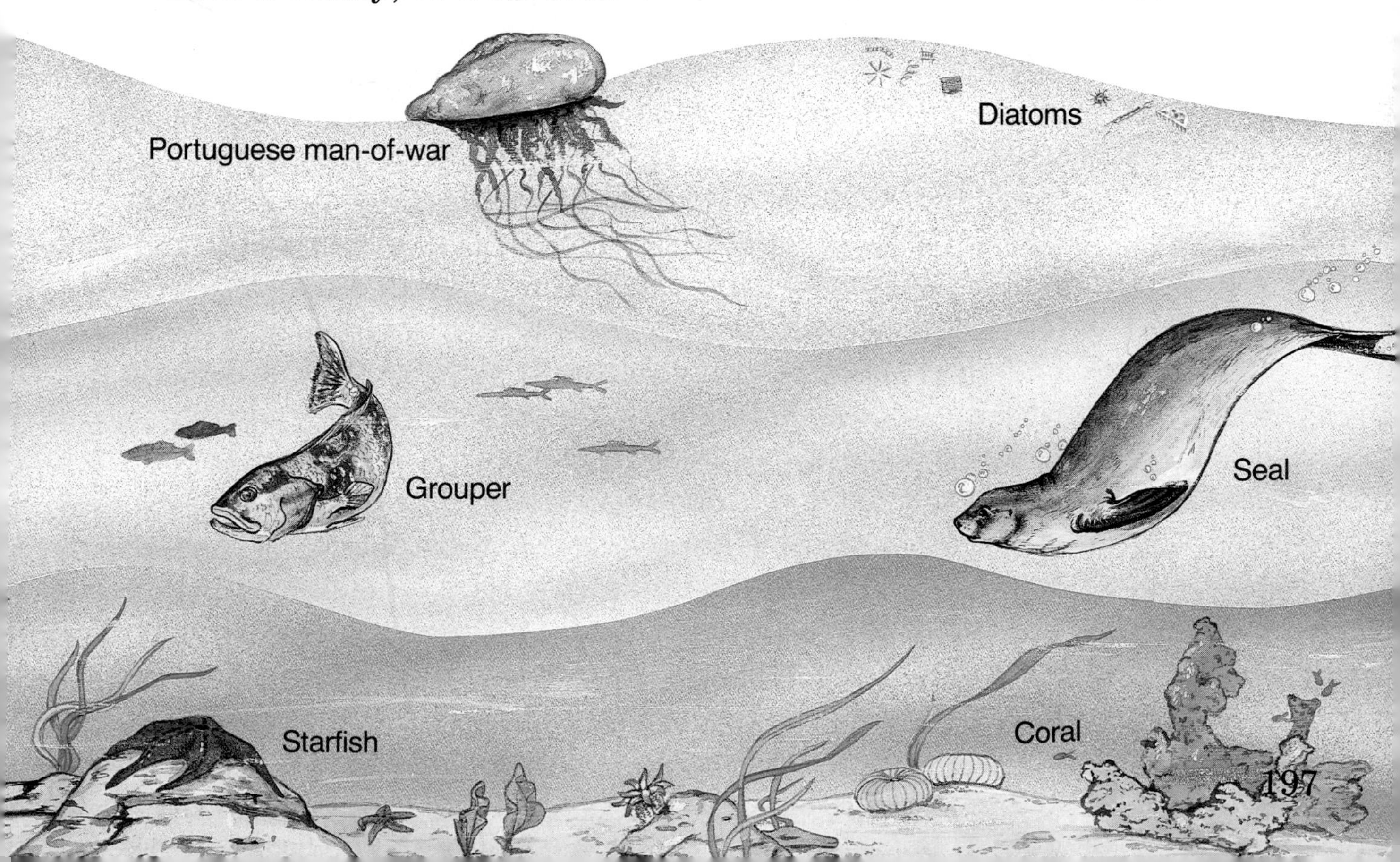

Kelp is an important natural resource.

Another kind of benthos is a brown seaweed called kelp. As explained in Chapter 8, kelp is a very important natural resource. It is used as an ingredient in foods such as ice cream and salad dressing. It is also an ingredient in certain medicines, cosmetics, and fertilizers.

Let's look more closely at how groups of ocean life are dependent on one another. Plankton provides the base of food for most ocean life. You remember that some of the plantlike plankton is eaten by the animallike plankton. Free-swimming fish also eat plankton as well as other smaller ocean animals. When animals die, they begin to sink. Dead animals are used for food by other animals that live in the lower depths of the ocean. Ocean animals also give off waste products that are changed, in part, to nutrients. Currents carry the nutrients back up to the surface to be used by the plankton for food. You can see how important ocean life is at each level. If you were the fishing guide for John's family, where would you go to fish?

ACTIVITY

You Can...

Make a Collage of Ocean Life

Draw or collect pictures of various forms of ocean life. Use old newspapers or magazines if you collect the pictures. Use various types of materials if you draw the pictures. Arrange the pictures on a large poster board according to their location in the ocean. How are ocean life forms different from one location to another?

Lesson Summary

- Living things are found at different depths of the ocean.
- Diatoms and crab larvae are examples of plankton. Sea turtles and whales are examples of nekton. Barnacles and corals are examples of benthos.
- Living things at each level of the ocean are important parts of food cycles.

Lesson Review

1. Compare the three types of ocean life.
2. Name two ocean organisms that provide important natural resources. What resources do they provide?
★3. Describe how each type of ocean life is part of ocean food cycles.

Protecting the Ocean

LESSON 3 GOALS
You will learn
- that we must protect our ocean resources.
- how we can protect our resources.

Mary is excited. Her father has promised to take the family to the beach on Saturday. She can't wait to build sand castles and look for shells. However, when they got to the beach on Saturday something was different. Instead of seeing people enjoying the beach, they saw people with large containers picking up dead fish that had washed ashore. Mary could tell from the way the people worked that something was not right. What do you think could have caused the fish to die?

Because our oceans are so important to us, we must protect them for the future. But that's not as easy as you might think. Oceans have pollution problems. For example, sometimes oil wells and ships leak oil into the oceans. **Oil spills** kill living things on or near the surface of the ocean. They can also kill living things that live on the shore. Oil spills are very hard and expensive to clean up.

Oil pollution

Wastewater pollution

Some of the garbage and wastewater we throw away each day ends up in the ocean. Materials like foam and plastic that can't be broken down into products that are not harmful cause a problem for ocean life. Wastewater from industry has harmful chemicals that can kill living things. Chemicals used by farmers run off into rivers and eventually find their way into the ocean. Warm water that some power plants pump back into the ocean is harmful to the animals that need cool water.

How is warm water from power plants harmful to ocean life?

You may be thinking there is nothing you can do to solve these problems of pollution. However, pollution is everybody's problem. You can become aware of the problems that will occur if we don't protect our oceans. Problems of pollution affect all living things in and out of the ocean. What are some ways you can help in regard to this problem?

How we use our resources now will have an effect on what resources we have in the future. For example, we can use products that decompose, or break down over time, and be careful about how we get rid of trash. Then we won't have to worry about harming living things in our environment. Everyone must help conserve the resources we have.

Recycling helps save our resources.

Lesson Summary

- We will have resources for future use if we protect our resources.
- Becoming aware of how to protect our resources is important.

Lesson Review

1. Give three examples of possible pollution problems within oceans.
2. What are two ways in which people can help reduce pollution of oceans?

★3. What reasons would you give in trying to convince someone to conserve or save natural resources?

Use Application Activity on pages 365, 366.

How can oil spills be cleaned up?

What you need

5 clear plastic cups
vegetable oil
cotton gauze
powdered detergent
feathers
filter paper
piece of screen
pencil and paper
spoon

What to do

1. Fill each plastic cup half full of water.
2. Place a spoonful of the vegetable oil in each cup.
3. Experiment with all of the materials supplied by your teacher. Determine which type of material would be most useful in cleaning up an oil spill on the ocean.
4. Record what you observe when you use each material.

What did you learn?

1. What happened to the feathers when they were placed on the oil spill?
2. How useful would feathers be to clean up oil spills?
3. List another type of material that might be useful in cleaning up oil spills. Explain how it might work.

Using what you learned

1. What could we do if natural feathers were not available for oil spill cleanup?
2. What problems would arise if a material used to clean the spill caused the oil to become more dense than water and sink to the ocean bottom?

CHAPTER REVIEW 9

Summary

Lesson 1
- The ocean is a valuable resource.
- Sonar is used to study the ocean floor.
- Features of the ocean floor include the continental shelf, continental slope, plain, rift zones, and trenches.

Lesson 2
- Living things are found at different depths of the ocean.
- Diatoms and crab larvae are examples of plankton. Sea turtles are examples of nekton. Barnacles are examples of benthos.
- Ocean organisms are important parts of food cycles.

Lesson 3
- If we protect our resources they will exist in the future.
- It is important to learn how to protect our resources.

Science Words

Fill in the blank with the correct word or words from the list.

sonar	**plain**
continental shelf	**rift zones**
continental slope	**trenches**
nekton	**plankton**
diatoms	**benthos**
oil spills	

1. The gently sloping part of land covered by ocean water is the ____.
2. Cracks in the ocean floor where large mountain chains are located are called ____.
3. The bottom of the deep, dark ocean is the ____.
4. The ____ has canyons with steep sides.
5. Reflected sound waves, or ____, are used to find out how deep the water is.

Questions

Recalling Ideas

Correctly complete each of the following sentences.

1. The most important area for fishing is the
   (a) continental slope.
   (b) continental shelf.
   (c) plain.
   (d) rift zones.
2. A source of oil and natural gas in the oceans is the
   (a) continental shelf.
   (b) continental slope.
   (c) plain.
   (d) rift zones.
3. Large free-swimming organisms are
   (a) plankton. (c) benthos.
   (b) nekton. (d) kelp.
4. An example of a benthos organism is
   (a) a jellyfish. (c) an oyster.
   (b) a turtle. (d) a diatom.
5. Organisms that drift and move along with the help of surface currents are
   (a) plankton. (c) benthos.
   (b) nekton. (d) corals.
6. Plants and animals that live on the ocean bottom are
   (a) plankton. (c) benthos.
   (b) nekton. (d) all of these

Understanding Ideas

Answer the following questions using complete sentences.

1. What are the three main parts of the ocean floor?
2. How do trenches form?
3. What are some ocean pollution problems?
4. What things can we do to protect our resources?

Thinking Critically

Think about what you have learned in this chapter. Answer the following questions using complete sentences.

1. Why are our oceans important to us?
2. How would the food for ocean life be affected if there were no ocean currents?

UNIT REVIEW 2

Checking for Understanding

Write a short answer for each question or statement.

1. Describe the structure of our solar system.
2. Describe the moon's orbit around Earth.
3. Compare the sun with other stars.
4. Explain what happens during a solar eclipse.
5. What are the major oceans of Earth?
6. How do the oceans affect our climate?
7. What natural resources are found in the ocean?
8. Describe the features of plankton, nekton, and benthos.
9. Describe the color and temperature of the sun.
10. Why is Pluto a cold, dark planet?
11. Why are some stars brighter than others?
12. Why is Mars called the "red" planet?
13. How does the sun produce energy?
14. Why can the corona be seen only during a solar eclipse?
15. What are sunspots and solar flares? How do they affect Earth?
16. Describe three kinds of ocean life and give an example of each.
17. Explain what causes tides.
18. How are waves produced?
19. What makes ocean water salty?
20. How does the temperature of ocean water vary with location? With depth?

Recalling Activities

Write a short paragraph for each question or statement.

1. What are moon phases?
2. How can you compare sizes of planets?
3. How does distance and size affect how bright a light looks?
4. What causes deep ocean currents?
5. Why is it easier to swim in the ocean?
6. How can you map the ocean bottom?
7. How can oil spills be cleaned up?

Project Ideas

1. Research space probes that have returned information about other planets. Draw several of the probes on a poster and identify their features.
2. Compare planets for conditions such as gravity, size, temperature, and so on. Create a poster that includes this information.
3. Build a wave tank to demonstrate the effects of wave action on different shorelines.

Books to Read

101 Questions and Answers About the Universe by Roy A. Gallant, Macmillan Publishers: New York, 1984.
Learn about the wonders of the universe.

Space Voyager by Wendy Boase, Simon and Schuster: New York, 1984.
Take an imaginary trip through the solar system.

The First Travel Guide to the Bottom of the Sea by Rhoda Blumberg, Lothrop, Lee and Shepard: New York, 1983.
Journey through the sea.

Physical Science

"The time has come," the Walrus said,
"To talk of many things:
Of shoes — and ships — and sealing wax —
Of cabbages — and kings —
And why the sea is boiling hot —
And whether pigs have wings."

from *The Walrus and the Carpenter*
Lewis Carroll

SCIENCE
In Your World

CHAPTER 10

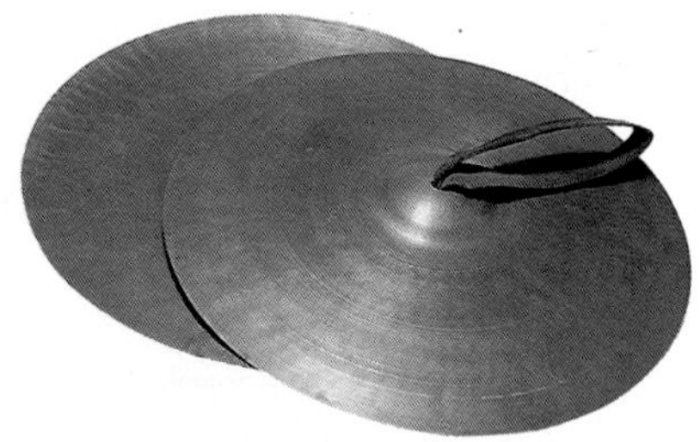

Light and Sound

Perhaps you have been lucky enough to hear the beautiful sounds of a xylophone or glockenspiel, or the rat-tat-tat of a drum. Although many different kinds of instruments make music, these belong to a special group called percussion instruments. They make music when talented musicians carefully strike their surfaces to cause vibrations.

Have You Ever...

Made Music With a Glass?

Using a metal spoon, lightly tap several glasses until you find one that makes a nice "ding" sound. Now fill the glass half full with water, and listen. Add more water until the glass is full and tap it again. What happens to the sound as the amount of water increases?

Light

LESSON 1 GOALS
You will learn
- that light has several characteristics.
- how light can be refracted or reflected.
- that different kinds of matter affect light differently.

When you think of light, what comes to your mind? Every day we use different kinds of light. Look around you. What kinds of light are you using right now? Where did the light come from? What would it be like if you didn't have this light?

We use light that comes from different sources. Some objects, like our sun and other stars, produce their own light. Other objects, such as light bulbs, produce light only when they are switched on.

Energy as Light

Light is one kind of energy. For an object to give off light, it must first take on extra energy. The extra energy may be provided by friction, heat, electricity, or many other forms of energy. Some of the extra energy is released from the object as light.

Extra energy released as light

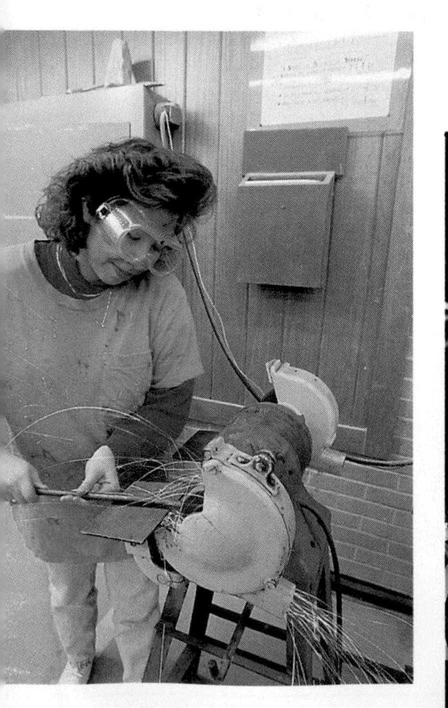

How do you think a light bulb makes light? When you turn on the light switch, electricity causes the filament inside the bulb to become very hot. The filament gives off extra energy as light.

Lightning is seen before thunder is heard.

Characteristics of Light

Light travels at a very high speed. In fact, it travels much faster than sound. A storm provides a good example of this. Have you noticed what happens when a storm is coming? If lightning is seen, you always see it before you hear thunder.

Light travels through most gases and liquids and through some solids. Because light can also travel through empty space, we can see objects such as the stars and moon in space.

Light reflected from a smooth surface and a rough surface

Reflection

When light strikes some objects, it bounces off. The bouncing back of light from a surface is called **reflection** (rih FLEK shun). Why can you see your reflection in a mirror? You see your image because the light that hits your body bounces off to hit the smooth surface of the mirror. The light then bounces back from the mirror in the same pattern. So you can see a reflection of your image.

How does light reflect from rough surfaces?

When light bounces back from a rough surface such as concrete, however, it scatters in different directions. If you were to look at a concrete wall, could you see a reflection of your image? In this case, you can't see a reflection because the light does not bounce back in the same pattern. It reflects in different directions.

Refraction

Light bends when it passes from one type of matter into another at an angle. This bending of light is called **refraction** (rih FRAK shun).

To better understand the refraction of light imagine you are skating along a sidewalk as shown in the pictures. As long as both of your skates are on the sidewalk, you will go straight. But if one skate hits the grass, you will turn into the grass. This is because the grass slows down only one skate. If you skate straight into the grass, you will be slowed down, but you won't turn because both skates will hit the grass at the same time.

SCIENCE AND . . .
Reading

Lightning was used in this lesson to show that—

A. sound travels faster than light.
B. light travels faster than sound.
C. sound can travel through space.
D. light travels faster in iron than in air.

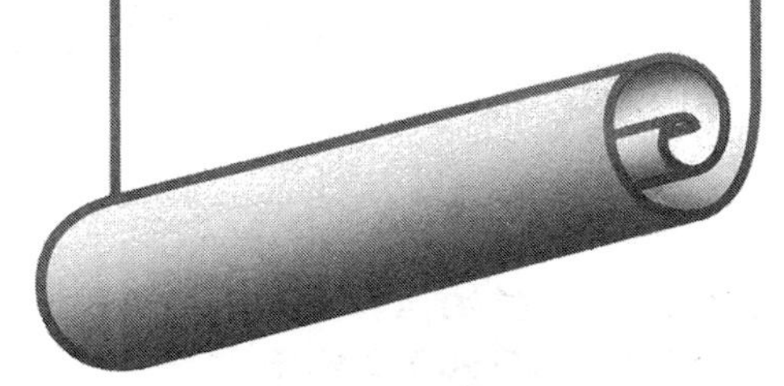

Maintaining speed and direction

Changing speed and direction

Changing speed

Light is refracted in much the same way. When light passes through different kinds of matter at an angle, it is bent. This is because the light changes speed as it travels through different types of matter.

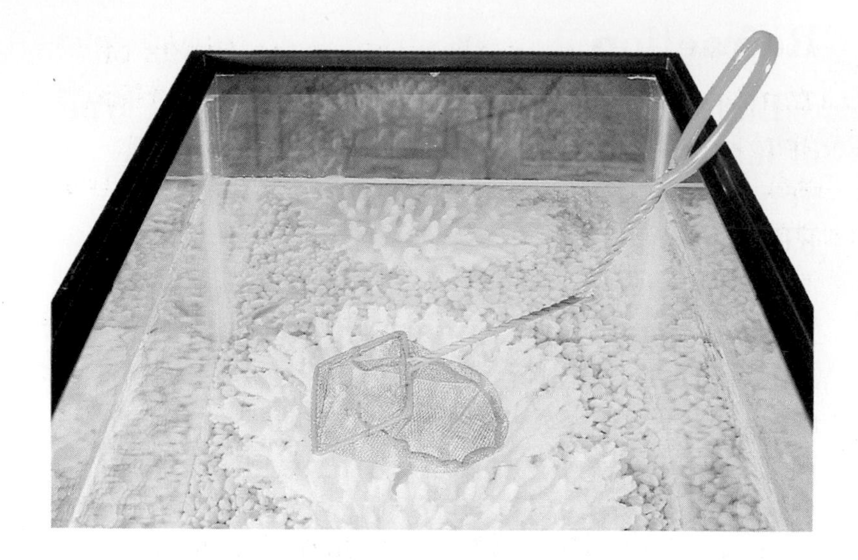

The handle appears bent because light travels from the water to the air.

The picture of the fishnet in the aquarium shows how light is refracted as it passes from water to air. Why does the handle of the fishnet appear to be bent? Where else have you seen the effects of refracted light?

Effects of Matter on Light

What is transparent matter?

Different kinds of matter affect how light passes through them. When light passes straight through **transparent** (trans PER unt) **matter,** we can see objects clearly. Most windows and eyeglass lenses are made of transparent glass or plastic. What are some other kinds of transparent matter?

Translucent (trans LEW sunt) **matter** lets light pass through, but it scatters the light in many directions. A lamp shade, frosted glass, and waxed paper are translucent. How do objects appear when seen through translucent matter? When light passes through this kind of matter, you can't see objects clearly.

Light can't pass through some kinds of matter at all. **Opaque** (oh PAYK) **matter** blocks light completely. Your classroom wall is opaque. Foil and cardboard are also opaque matter. What are some other examples of opaque matter?

In this lesson you have learned that we depend on light from many sources each day. You use light from the sun and other sources just to see where you are going. You also use light to help you see the size, shape, and color of things around you. Light helps you know more about the world around you.

Transparent, translucent, and opaque wrappings

ACTIVITY

You Can...

Bend a Pencil!

Get a clear glass and put a pencil in it as shown. Look down at the pencil from an angle. Draw a picture of what you see. Now fill the glass about half full with water. Look at the pencil again. Draw a picture of it now. What happens to how the pencil looks? Why does this happen?

Lesson Summary

- Light travels faster than sound and can travel through most gases and liquids, through some solids, and through empty space.
- When light bounces off an object, it is reflected. When light is bent when passing from one type of matter into another, it is refracted.
- Transparent, translucent, and opaque objects affect light differently.

Lesson Review

1. What is needed before an object can give off light?
2. Waxed paper is an example of what kind of matter?

★3. What is the difference between reflection and refraction?

Use Application Activity on pages 367, 368.

How can you bend light?

What you need
butcher paper
clear, rectangular pan
metric ruler
marking pen
pencil and paper
water

What to do
1. Trace the outline of the pan on the paper.
2. Remove the pan. Use a ruler and draw **Line 1** straight through the outline as shown.
3. Starting from the same point, draw **Lines 2, 3,** and **4** as shown.
4. Put the pan back on the paper. Add water to a depth of 5 cm.

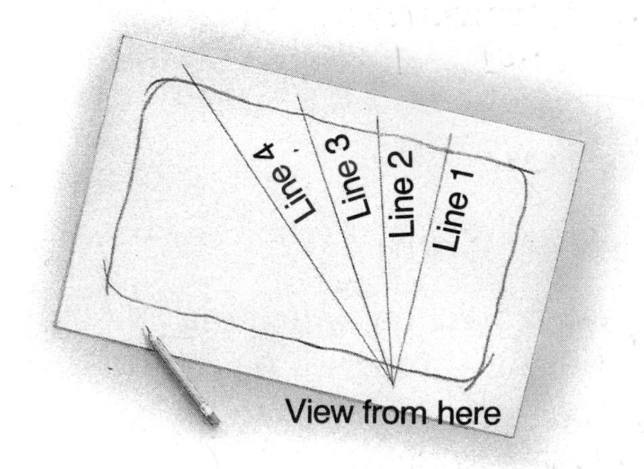

5. Look through the water along **Line 1** as shown. Hold your pencil behind the pan and try to line it up along **Line 1.** Mark this point on the paper.
6. Repeat step 5 for **Lines 2, 3,** and **4.**
7. Remove the pan of water and measure how far away from the lines your pencil marks are. Record your results and observations.

What did you learn?
1. How close did you place your pencil to **Line 1?** To **Lines 2, 3,** and **4?**
2. What caused you to miss the lines?

Using what you learned
1. As you go from **Line 2** to **4,** what happens to the distance your mark was from the line? Why?
2. Why should you know light bends when you go fishing?

Color

LESSON 2 GOALS
You will learn
- that white light is made up of many colors.
- how a prism separates white light into the visible spectrum.
- why objects appear certain colors.

Colors are useful and important to us in many ways. Have you ever tried to find a friend among a large group of people? You probably look first for people who are wearing the same color clothing as your friend, and then look for more details. You also use the colors of a traffic light to know when it is safe to cross a busy street. Do you know why we see colors as we do? In this lesson, you will learn how light is affected by matter to show colors.

Color has importance in daily life.

A Rainbow of Colors

Think about when you have seen the bright sun shining through a piece of crystal. You may have seen a small band of colors on the wall or floor.

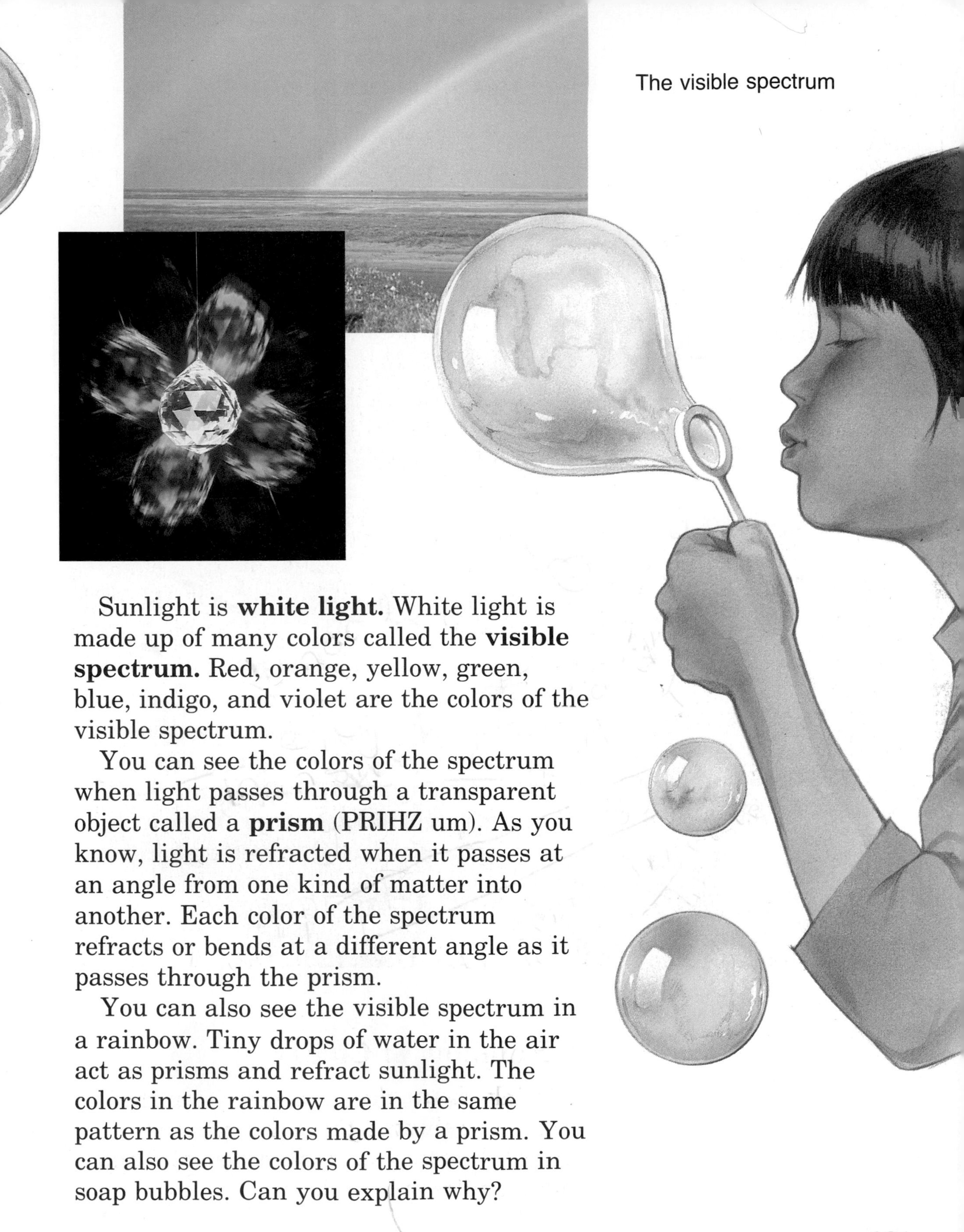

The visible spectrum

Sunlight is **white light.** White light is made up of many colors called the **visible spectrum.** Red, orange, yellow, green, blue, indigo, and violet are the colors of the visible spectrum.

You can see the colors of the spectrum when light passes through a transparent object called a **prism** (PRIHZ um). As you know, light is refracted when it passes at an angle from one kind of matter into another. Each color of the spectrum refracts or bends at a different angle as it passes through the prism.

You can also see the visible spectrum in a rainbow. Tiny drops of water in the air act as prisms and refract sunlight. The colors in the rainbow are in the same pattern as the colors made by a prism. You can also see the colors of the spectrum in soap bubbles. Can you explain why?

Seeing Individual Colors

We see color that is reflected.

We see objects when they reflect light to our eyes. If an object appears blue, it reflects blue to our eyes and absorbs all other colors of the spectrum. Red apples look red because they reflect red and absorb all other colors.

The zebra in the picture has both black and white hair. The white hair looks white because it reflects all the colors of the visible spectrum. The black hair looks black because it absorbs almost all the colors of the visible spectrum. Very little visible light is reflected to our eyes from black surfaces.

African zebra

In this lesson, you learned that an object appears a certain color because it reflects light of that color to your eyes. Other colors of light are absorbed by that object. Because you can see color you can tell a red object from a green or yellow one.

Being able to see color is very important. What are some reasons why color is important to you?

Lesson Summary

- White light is made up of all the colors called the visible spectrum.
- A prism is a transparent object that refracts white light and separates it into the colors of the spectrum.
- Colored objects reflect certain colors of the spectrum and absorb the rest.

Lesson Review

1. What are the colors of the visible spectrum?
2. Why can we see the visible spectrum in a rainbow?

★3. What colors of light are absorbed by an object that appears green?

Sound

LESSON 3 GOALS
You will learn
- how vibrations cause sounds that travel through solids, liquids, and gases.
- what the properties of sound are.
- how sound waves can be reflected or absorbed.

There are sounds around you all the time. What sounds can you hear in your classroom right now? Do you know what causes these sounds?

Sounds are always around us.

Sound is caused by **vibrations** (vi BRAY shunz). When an object vibrates back and forth, it pushes on small particles of air, or air molecules, causing the air to vibrate. These moving molecules bump into other nearby air molecules, causing them to vibrate, too. In this way, sound vibrations move through air or other matter in the form of sound waves. The sound waves vibrate our ear drums, enabling us to hear the energy of the original vibrating object as sound. From this example we can understand that sound is a form of energy because it causes matter to move.

Sounds are caused by vibrations.

Sometimes you can see objects vibrate. When you strum a guitar string, it moves back and forth as it vibrates. The vibrations of the guitar string create sound waves. These sound waves travel through the air to your ears.

You can also feel some vibrations. The sound of your voice is caused by vibrations of your vocal chords. You can feel the vibrations that make your voice if you place your fingers against your throat while you talk.

Through what kinds of matter can sound travel?

Sound waves travel through different kinds of matter—gases, solids, and liquids. You know sound travels through air because you can hear noises in your room. If you listen closely, you can probably hear sounds coming through solid walls from the classroom next door. If you swim underwater, you know that you can hear sounds that travel through liquids.

Because sound waves are caused by vibrations of particles of matter, they travel only through matter. Sound waves can't travel through empty space such as a vacuum. A vacuum (VAK yewm) is space that has no matter.

Sound waves travel at different speeds in different kinds of matter. You can look at the table to see how far sound waves travel in one second through different kinds of matter.

Table 1 Speed of Sound Through Matter

Matter at 20°C	State	Meters per second
Air	Gas	340
Water	Liquid	1,500
Brick	Solid	3,600
Wood	Solid	3,800
Iron	Solid	5,200

Properties of Sound

What is volume?

Sounds can be loud or soft. The loudness or softness of a sound is called **volume** (VAHL yum). Loud sounds have more volume and soft sounds have less volume. Larger amounts of energy produce louder sounds. Less energy produces softer sounds. For this reason, loud sounds transfer more energy than soft sounds do.

Instruments can produce sounds of different volume, pitch, and frequency.

Not only can sounds differ in their volume, they also can differ in their pitch. **Pitch** is the highness or lowness of a sound. Objects that vibrate very fast make high-pitched sounds. Objects that vibrate more slowly make low-pitched sounds.

Why does a sound have a low pitch?

The number of times an object vibrates in one second is called its **frequency** (FREE kwun see). An object that vibrates many times in one second has a high frequency and will make a high-pitched sound. An object that vibrates few times in one second has a low frequency and makes a low-pitched sound.

When strummed, a guitar's thicker strings vibrate more slowly than its thin strings. The thicker strings produce sounds of lower pitch. A guitar player can make each string produce a higher pitch by pressing on the string while strumming. The shorter string will vibrate faster, producing a higher-pitched sound.

A bullwhip's crack is produced when the tip breaks the speed of sound, creating a small sonic boom.

You learned that light is reflected when it bounces off smooth surfaces. Sound waves react in much the same way when they bounce off or are reflected by some kinds of matter. An **echo** is one kind of reflected sound. In an echo, sound waves bounce off a surface and are reflected back toward their source. Smooth, hard surfaces, like the walls and floor of a gym, reflect sound best. Where are places you've heard an echo of your voice?

Not all matter reflects sound waves equally well. Soft or rough surfaces absorb more sound waves than do hard, smooth surfaces. Look at the pictures. One room has carpet and drapes while the other room has a wood floor and no drapes. In which place would sound be absorbed better? Why do you think so?

Furnishings within rooms affect sound.

Sounds may be pleasant or unpleasant, loud or soft, high or low. Some sounds warn you of danger, while others let you communicate with someone else. How do sounds make your world more interesting?

Lesson Summary

- Sound travels when a vibrating object causes molecules around it to also vibrate.
- Volume, pitch, and frequency are properties of sound.
- Sound waves can be reflected by hard, smooth surfaces or absorbed by soft or rough surfaces.

Lesson Review

1. Why is sound considered a form of energy?
2. What kind of sound is made by high-frequency vibrations?

★3. What is an echo?

Use Application Activity on pages 369, 370.

What do sound vibrations look like?

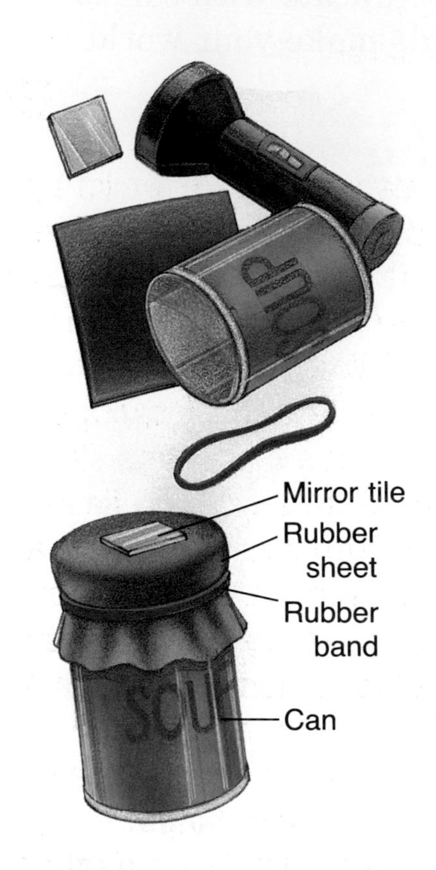

What you need

small mirror tile
paste
rubber sheet
tin can (both ends removed)
heavy rubber band
flashlight
drawing paper
pencil and paper

What to do

1. Stretch the rubber sheet over one end of the can and hold it in place with the rubber band.
2. Carefully paste the mirror to the rubber sheet and set it aside to dry. *The mirror should be slightly off center.*
3. Darken the room. Have a partner shine the flashlight on the mirror. Observe the reflection on the wall.
4. Sing a musical scale into the open end of the can. Draw a picture of the reflection.
5. Sing loudly, then softly into the can. Draw the reflections.

What did you learn?

1. Compare the pictures of the reflections produced when you:
   a. sang a musical scale.
   b. sang softly, then loudly.
2. Which sound produced the most unusual pattern?

Using what you learned

1. What caused the patterns you saw in the reflections?
2. How did the pictures of your reflections compare with others?

I WANT TO KNOW ABOUT...

Biometrics

A security guard walks up to the door of a top secret area. A pair of binoculars are mounted next to the door. The guard looks into them for a few seconds. The door swings open. Does this sound like a scene from a science fiction movie? An eye scanner is a kind of security system that is being used today! This system is called a biometric (bi oh ME trihk) security system.

The word *biometric* means to measure living things. Biometrics differs from other security systems in one important way. Most security systems check for some special "thing" like an entrance card or a key. Biometrics checks you against earlier measurements made from your body.

There are four main types of biometric systems. These systems may measure fingerprints, voice, palm of the hand, or vein patterns in the eye. The patterns made from fingerprints, voice, palm, or eye are different for each person. It is difficult to make false copies of these patterns.

One big problem with the biometric systems is that they are very expensive. Scientists are working to build less expensive systems. Some day biometrics may be used to protect your home.

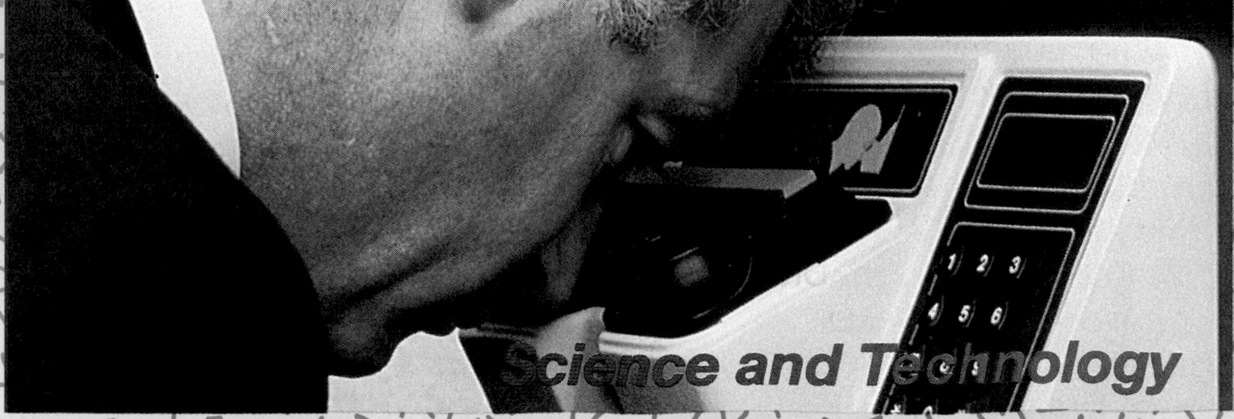

CHAPTER REVIEW

10

Summary

Lesson 1

- Light can travel through all types of matter and space.
- When light bounces off an object, it is reflected. When light is bent when passing from one type of matter into another, it is refracted.
- Transparent, translucent, and opaque objects affect light differently.

Lesson 2

- White light is made up of all the colors of the spectrum.
- A prism is a transparent object that refracts white light and separates it into the colors of the spectrum.
- Colored objects reflect certain colors and absorb the rest.

Lesson 3

- Sound is caused by vibrating molecules. Sound travels in waves.
- Volume, pitch, and frequency are properties of sound.
- Sound waves can be reflected or absorbed.

Science Words

Fill in the blank with the correct word or words from the list.

reflection	**visible**
refraction	**spectrum**
transparent matter	**prism**
translucent matter	**vibration**
opaque matter	**volume**
white light	**pitch**
	frequency
	echo

1. One kind of reflected sound wave is a(n) ____.
2. Light can't pass through ____.
3. Light can be clearly seen when it passes through ____.
4. The bouncing back of light from a surface is called ____.
5. White light is made up of many colors called the ____.
6. Sunlight is ____.

7. The bending of light as it passes from one type of matter into another is called ____.
8. A(n) ____ is a transparent object that refracts each color of the spectrum at different angles.
9. Sound is caused by a(n) ____.
10. The highness or lowness of a sound is caused by its ____.

Questions

Recalling Ideas

Correctly complete each of the following sentences.

1. Light can travel through
   (a) space. (c) gases.
   (b) liquids. (d) all of these.
2. Vibrations cause
   (a) sound. (c) the spectrum.
   (b) light. (d) reflections.
3. Space that has no matter is
   (a) a vacuum. (c) a liquid.
   (b) a solid. (d) all of these.
4. The loudness or softness of a sound is called
   (a) pitch. (c) frequency.
   (b) volume. (d) an echo.
5. The number of times an object vibrates in one second is its
   (a) pitch. (c) volume.
   (b) frequency. (d) echo.

Understanding Ideas

Answer the following questions using complete sentences.

1. Why is light refracted as it travels through different kinds of matter?
2. Compare the speed of light to the speed of sound.
3. What are the colors of the visible spectrum?
4. What are the properties of sound?

Thinking Critically

Think about what you have learned in this chapter. Answer the following question using complete sentences.

1. What colors of light does a red blanket reflect and absorb?

SCIENCE
In Your World

Electricity

Static electricity can make your hair stand on end! You may have had a similar hair-raising experience by combing your hair on a dry day. Actually, this will happen whenever your hair loses electrons and becomes positively charged.

Have You Ever...

Seen Bashful Balloons?

Blow air into two balloons and tie a 30-centimeter string to each of them. Rub both balloons against your hair to create an electrical charge. Then hold the strings to suspend the balloons from one hand. Why are they so bashful? Now rub one balloon against your hair and place it on the wall. Is it still bashful?

Static Electricity

LESSON 1 GOALS
You will learn
- that an atom has three parts.
- about static electricity.
- what causes a static discharge.

When Mary walked across the carpeted floor and touched the doorknob, she felt a shock. In another room, her little sister wants to know why the laundry makes a crackling sound when their mother takes clothes out of the dryer.

Mary and her mother and sister are finding out about static electricity (STAT ihk • ih lek TRIHS ut ee). To understand static electricity, we must first learn about atoms.

Static electricity affects us at home.

An **atom** is the smallest part of matter. Look at the picture to see a model of an atom. The center of the atom is the **nucleus** (NEW klee us). The nucleus has two kinds of particles. One kind of particle is a **proton** (PROH tahn) and the other is a **neutron** (NEW trahn). A third kind of particle, called an **electron** (ih LEK trahn), is found outside the nucleus. Electrons form a cloud outside the nucleus.

Protons
Neutrons
Nucleus
Electrons

What are three parts of an atom?

Protons and electrons contain a small amount of electricity or electric charge. Protons have a positive charge (+) and electrons have a negative charge (−). Usually, the number of protons and electrons in an atom is the same, and the atom as a whole has no electric charge.

In many materials, however, some electrons can move from atoms of one material to atoms of a different material. When this happens, atoms of the material that loses electrons are left with a positive charge. Atoms of the material that gains electrons take on a negative charge.

What happens when you rub a balloon against a wool sweater? Some of the electrons move from the atoms of the wool sweater to the atoms of the balloon. The atoms of the balloon now have more electrons than protons, and take on a negative charge. The atoms of the wool sweater have more protons than electrons, and therefore have a positive charge.

SCIENCE AND . . .
Writing

Write a composition on "Lightning in My Clothes Dryer." Describe how and why static electricity works. Then, elaborate on the idea that static cling and lightning are closely related.

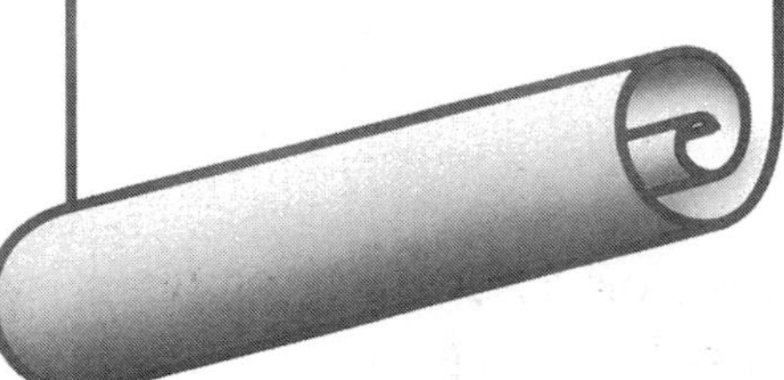

Electrons can move from one item to another.

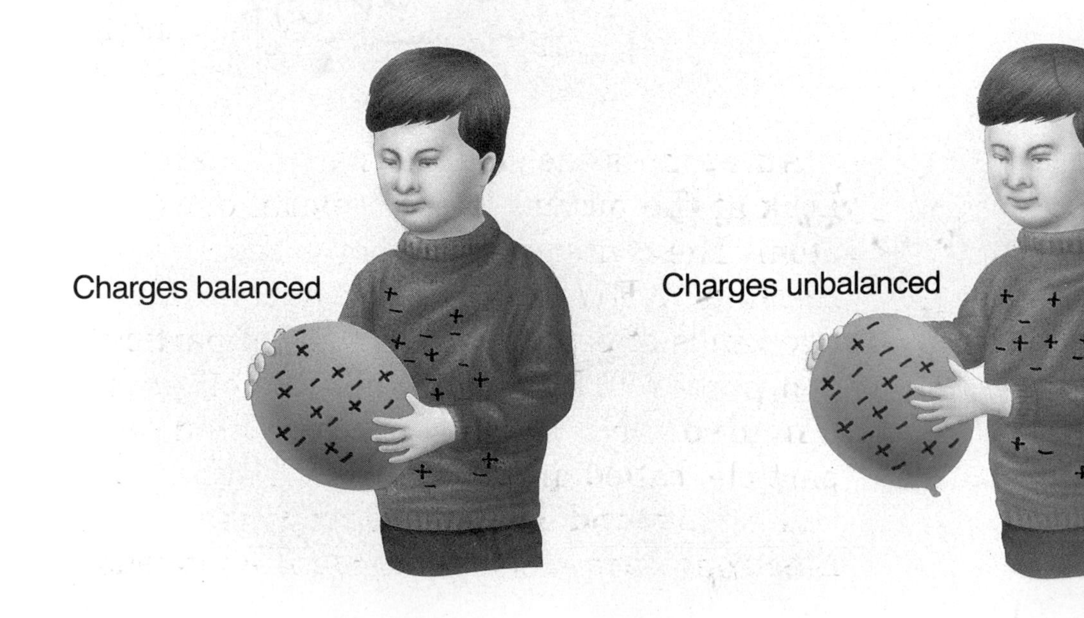

When the number of protons and electrons is unequal, static electricity is produced. **Static electricity** is the charge on an object that has an unequal number of protons and electrons.

What is static electricity?

You may have noticed an effect of static electricity while combing your hair. When you comb your hair, electrons from your hair can be rubbed off onto the comb. The comb has a negative charge because it has more electrons than protons. What about your hair? Since your hair lost electrons, it has a positive electric charge.

Objects with electric charges may attract or repel other objects. If objects have opposite charges, they attract, or pull toward, each other. But if objects have the same charge, they repel, or push away from, each other. Your hair and the comb are attracted to each other because they have opposite charges. However, because each hair has the same electric charge, the hairs repel each other.

You may notice static electricity when you comb your hair.

Now we can explain why clothes in a dryer cling together. While being tumbled together in a dryer, clothes rub against each other. Electrons move from one kind of clothes to another. Because opposite charges attract, the clothes cling together.

Static discharge

Sometimes electrons "jump" from one object to another object. When the electrons jump to another object, they cause a spark. Sometimes you may see the spark or hear a crackling sound. You can feel an electric shock when charges jump to or from your body. This happened to Mary as she touched the doorknob. When electrons move from one object to another, they cause static discharge. After static discharge, an object no longer has a positive or negative charge. The flow of electrons during the static discharge has brought the normal number of electrons to the atoms of each object.

Would You Believe?

Lightning is more likely to strike twice in the same place since electricity tends to follow the path of least resistance.

Lightning is another example of static discharge. Electrons can build up on clouds under certain weather conditions. When the negative charge builds up enough, the electrons jump to another object, such as another cloud or Earth. When this happens, we see a giant spark called lightning.

Lesson Summary

- The parts of an atom are protons, neutrons, and electrons.
- Static electricity is the charge on an object that has an unequal number of protons and electrons.
- Static discharge happens when electrons move from one object to another.

Lesson Review

1. Tell how protons, neutrons, and electrons are positioned within atoms.
2. Describe how static electricity is produced when hair is combed.

★3. How is lightning an example of static discharge?

What is static electricity?

What you need

2 balloons
masking tape
string
scissors
wool cloth
pencil and paper

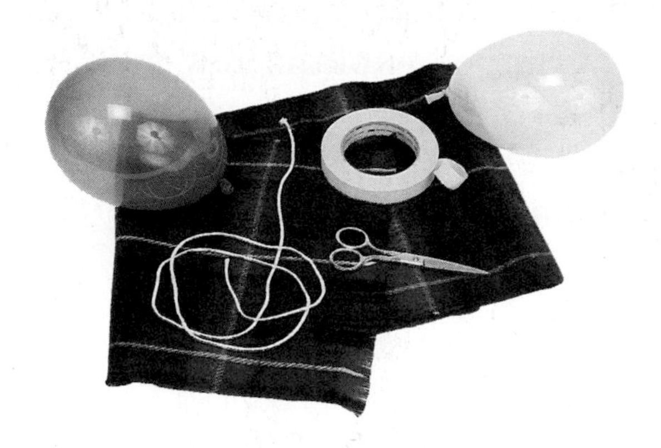

What to do

1. Blow up 2 balloons and tie the ends. Tie a piece of string to each balloon.
2. Tape the strings to your desk as shown in the picture.
3. Rub each balloon with a piece of wool. Observe what happens.
4. Draw a picture of the position of the balloons before and after you rubbed them with wool.

What did learn?

1. What happened to the balloons when you rubbed them with wool?
2. Where else have you seen this effect of static electricity?

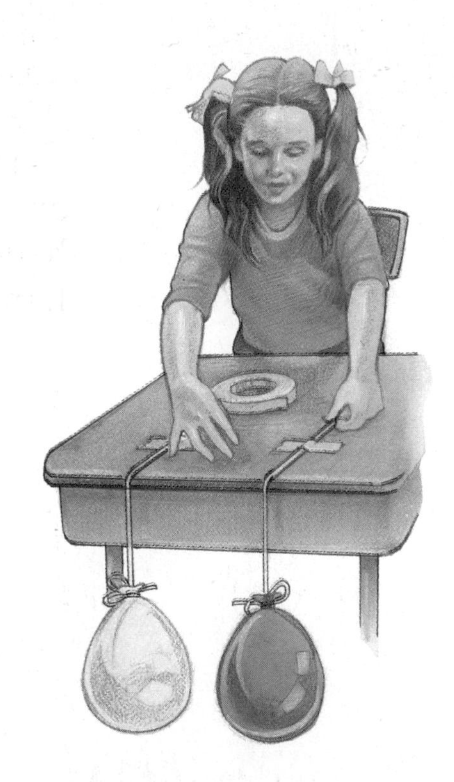

Using what you learned

1. How can you explain what happened to the balloons?
2. What do you think would happen if you touched each balloon after you rubbed it? Try it.

Current Electricity

LESSON 2 GOALS
You will learn
- about electric currents and circuits.
- how electrons move through a circuit.
- how insulators and conductors are different.

Bill was overjoyed. He had received a toy car for his birthday. This was the kind of car that could go forward or backward or around in circles. After carefully removing the car from the package, Bill wanted to see the car move on its own. He turned the control to *on* but nothing happened. He tried turning the wheels, but the car still wouldn't move on its own. Then Bill read the label on the package: BATTERY NOT INCLUDED. Now Bill knew why the car wouldn't work. The car needed a battery to move it along. Can you explain how a battery will cause Bill's car to move? In this lesson you will learn how a battery can be used to make electric current to make Bill's car run.

Bill wondered why the car wouldn't work.

You learned in Lesson 1 that atoms have protons and electrons. The protons in an atom don't move, but some of the electrons can move from atom to atom.

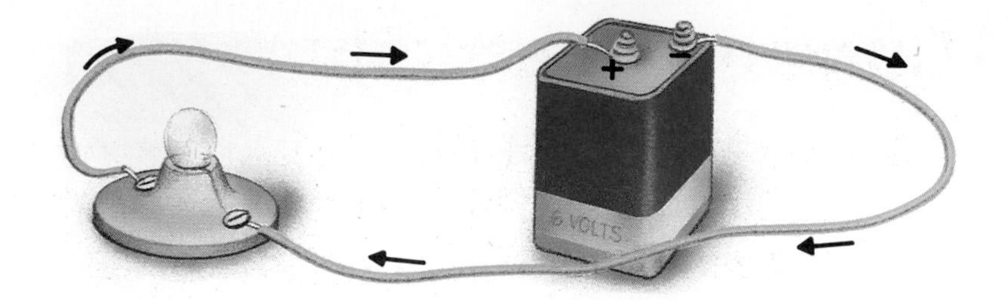

Electric current flows from negative to positive.

Electrons, which have negative charges, flow toward protons, which have positive charges. This movement of electrons along a path is called a **current.** An electric current is made only of flowing electrons. Electric current flows from negative to positive.

You can compare an electric current to a stream of water flowing downhill. Electric current follows a path in much the same way as the water in a stream follows a path. The path through which a current flows is called a **circuit** (SUR kut).

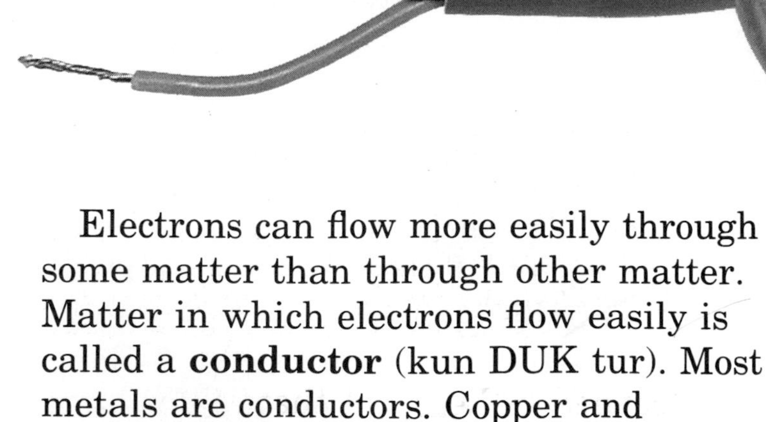

Electrons can flow more easily through some matter than through other matter. Matter in which electrons flow easily is called a **conductor** (kun DUK tur). Most metals are conductors. Copper and aluminum are metals that are used to make wire because they are good conductors.

ACTIVITY

You Can...

Find the Conductors

Connect the battery, wire, bulb, and bulb holder as shown. Touch the two loose wire ends together to make sure the bulb lights. Touch the ends of each wire to a paper clip. What happens? Predict which objects will light the bulb. Test your predictions. Which objects are conductors? What materials make the best conductors?

What different kinds of matter are insulators?

Matter in which electrons don't flow easily is called an **insulator** (IHN suh layt ur). Glass, plastic, wood, and rubber are insulators. Plastic and rubber are often used to cover wires. Why are insulators used to make these coverings on wires?

A **battery** can produce an electric current. Look at the picture of the battery. The path from the battery through the wire to the light bulb and back to the battery is a circuit. Electrons flow through the circuit to make the bulb light.

Circuits are like toy train tracks. As long as all of the train track is connected properly, the train can move along. But what happens if part of the track is missing or disconnected?

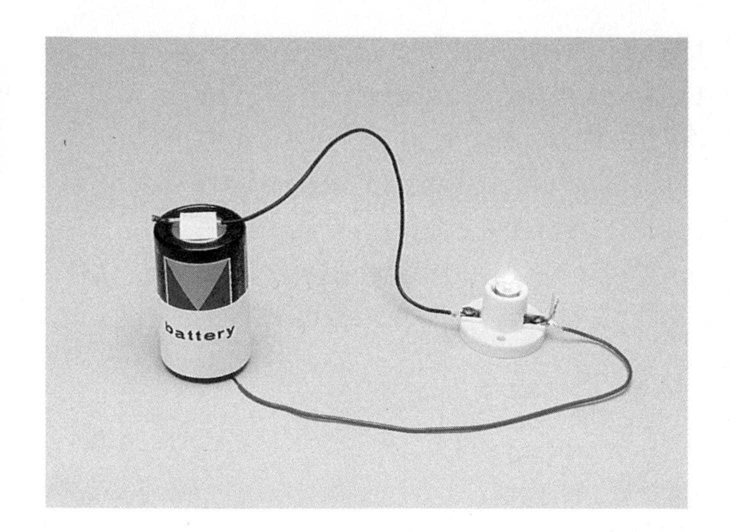

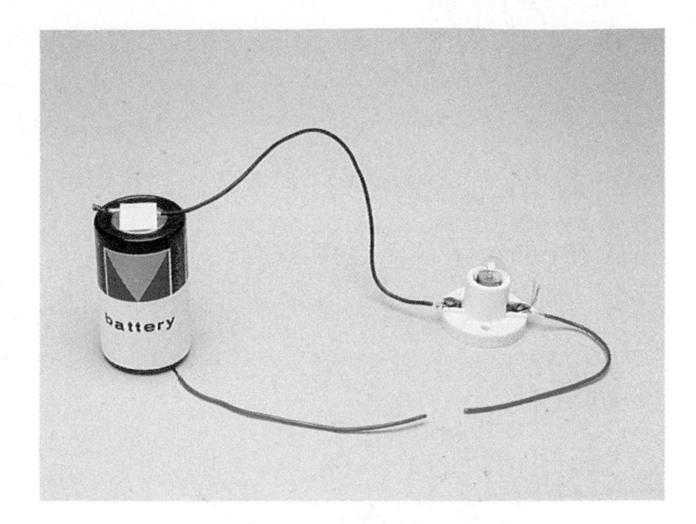

A closed circuit and an open circuit

Look at the pictures of the two electric circuits. In the picture on the left, current can flow from the battery through the wire to the bulb and back to the battery. The circuit is complete, or closed. The bulb will light.

Now look at the picture on the right. Will the bulb light? The circuit is incomplete, or open. The electrons don't have a complete path to flow through.

How is a switch used?

You close a circuit every time you turn a light on. You open a circuit when you turn a light off. You use a **switch** to open or close a circuit. A switch is made of material that is a conductor, but the handle of the switch is an insulator.

You can think of a stream of water coming into your house as a circuit. When you turn on the faucet, water flows freely from the pipes into the sink. This represents a closed electric circuit. Now, suppose you turn off the faucet. Water stops flowing into the sink. This represents an open electric circuit. The faucet is the "switch" used to open and close the circuit.

Flowing water represents a closed electric circuit.

Look at the picture of closed and open circuits with switches. The switch is the conductor. When the handle of the switch is down, there is a complete circuit. The circuit is closed. What happens to the circuit when the handle of the switch is up?

A switch can close or open a circuit.

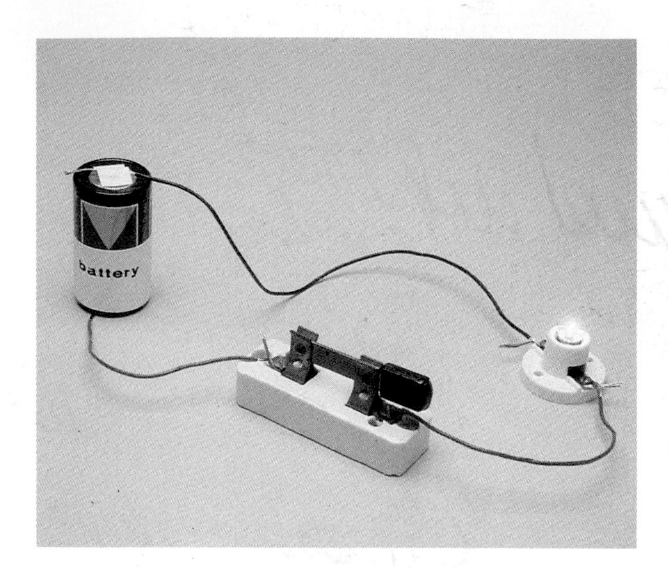

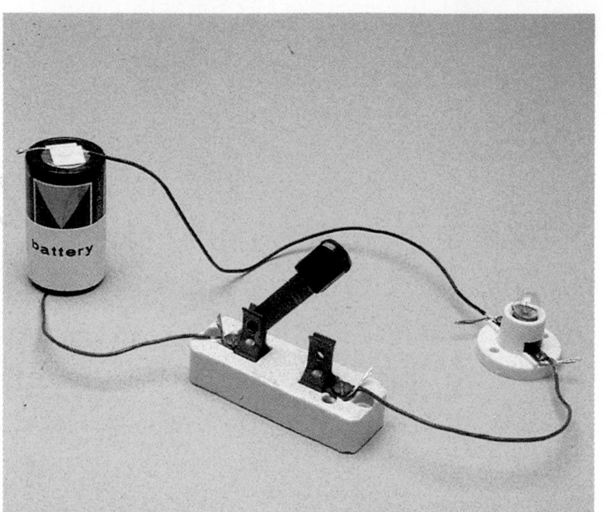

The light switches in your home or school work the same way. When the switch is on, a conductor in the switch connects two wires so that current can flow through the lights. When the switch is off, the circuit is open. The current can't flow through the lights.

Electric current is used to produce light and heat. You can see how light and heat are produced in your home and school every day. Electric current can also be used to cause motion or to make a motor turn.

At the beginning of this lesson, we read about Bill with his toy car. Bill bought a battery and put it into the toy car. Now when Bill turns on the control, what will probably happen?

The battery provides the energy that makes electrons flow. Electrons flow from the battery through the wires to the motor and back to the battery. The car moves if the switch is on. If the switch is off, current stops flowing through the circuit and the car won't move.

Lesson Summary

- Current is the movement of electrons along a conductor in a path called a circuit.
- A switch can be used to control the movement of electrons through a circuit.
- Electrons flow easily through conductors, but don't flow easily through insulators.

Lesson Review

1. How do conductors and insulators differ?
2. How is a switch used in a circuit?

★3. Sometimes glass covers are used at the top of utility poles that carry electric current. These glass objects enclose the electric connections from one wire to another. Why is glass used instead of a material such as copper or steel?

How can you make the bulb light?

What you need

2 flashlight batteries
2 flashlight bulbs
4 pieces of bell wire (15 cm each)
masking tape
pencil and paper

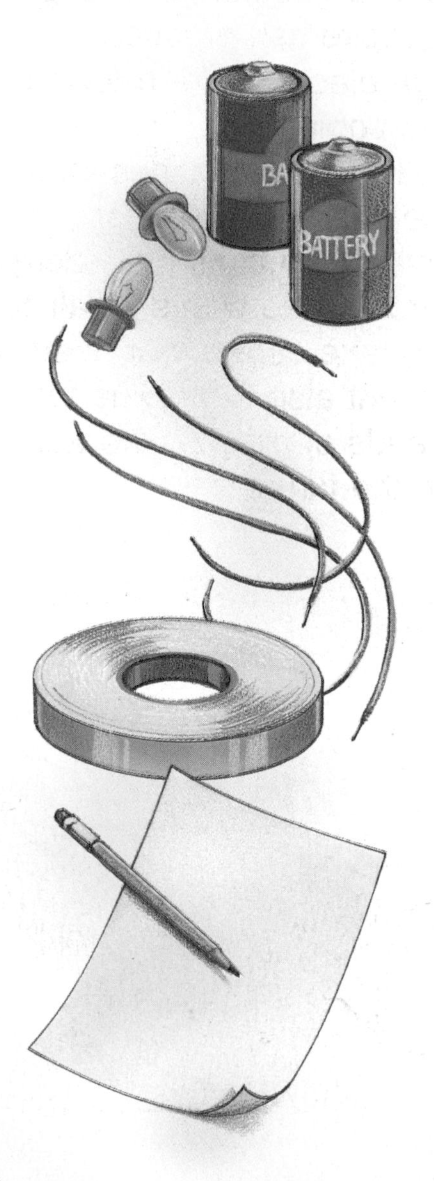

What to do

1. Using 1 battery and 2 wires, make the bulb light. Use the masking tape to hold the wires in place.
2. Draw a picture to show the circuit you made.
3. Light the bulb using 2 batteries. Draw the circuit you made.
4. Try to light both bulbs at once. Draw the circuit you made.

What did you learn?

1. How did you make the bulb light each time?
2. What supplied electric current to light the bulbs?
3. What difference was there when you used two batteries?

Using what you learned

1. Why was there a difference when two batteries were used to light the bulb?
2. What would the difference be if one battery and two bulbs were used? Why?

I WANT TO KNOW ABOUT...

Public Speaking

Phyllis Miranda works for an electric company. She speaks to school children and other groups about energy. Sometimes she talks about finding new resources for energy. Geothermal (jee oh THUR mul) energy is her newest topic.

To get geothermal energy, a hole is drilled into hot, dry rocks deep inside Earth and water is pumped into the hole. The rocks heat the water and change it into steam. The steam is pumped out of the hole and used to drive turbines at power plants.

Phyllis explains that a hole was drilled at a test power plant. The geothermal energy at the test site can produce enough electricity for a town of 4,000 people.

Phyllis Miranda hopes that people will learn from her speeches and conserve energy resources. She wants us all not only to take care of our needs today, but also to provide for the needs of people who will live in the future.

CHAPTER REVIEW

11

Summary

Lesson 1

- Atoms contain protons, neutrons, and electrons.
- Static electricity is the charge on an object that has an unequal number of protons and electrons.
- Static discharge happens when electrons move from one object to another.

Lesson 2

- Current is the movement of electrons along a conductor in a path called a circuit.
- A switch can be used to control the movement of electrons through a circuit.
- Electrons flow easily through conductors, but don't flow easily through insulators.

Science Words

Fill in the blank with the correct word or words from the list.

nucleus	**current**
atom	**proton**
neutron	**electron**
circuit	**conductor**
battery	**insulator**
static electricity	**switch**

1. A particle in the nucleus of an atom with a positive charge is a(n) ___ .
2. A(n) ___ is found outside the nucleus of an atom and has a negative charge.
3. Copper is a good ___ .
4. An atom's center is the ___ .
5. The path through which a current flows is a(n) ___ .
6. You use a(n) ___ to open or close a circuit.
7. The movement of electrons along a path is a(n) ___ .
8. ___ may affect hair when it is combed.
9. An atom's nucleus contains a proton and a(n) ___ .

Questions

Recalling Ideas

Correctly complete each of the following sentences.

1. The smallest part of matter is
   (a) an electron.
   (b) a proton.
   (c) an atom.
   (d) a neutron.
2. The part of an atom that can move to another atom is a(n)
   (a) proton. (c) neutron.
   (b) electron. (d) nucleus.
3. An electric current is made of flowing
   (a) protons. (c) neutrons.
   (b) electrons.(d) nucleus.
4. Most conductors are
   (a) metal. (c) wood.
   (b) glass. (d) plastic.
5. Most insulators are
   (a) glass. (c) wood.
   (b) plastic. (d) all of these

Understanding Ideas

Answer the following questions using complete sentences.

1. Describe an atom's structure.
2. What are the two kinds of electric charge?
3. How can an atom become positively charged?
4. How is static electricity produced?
5. What happens when two objects with opposite charges are brought together?
6. How is a static discharge produced?
7. In what direction do electrons flow?
8. What happens when objects with the same charge are brought together?

Thinking Critically

Think about what you have learned in this chapter. Answer the following questions.

1. Explain the flow of electrons and charge on each object as you walk across a carpeted room and touch a doorknob.
2. Think of a battery-operated toy. Explain how the batteries make it work.

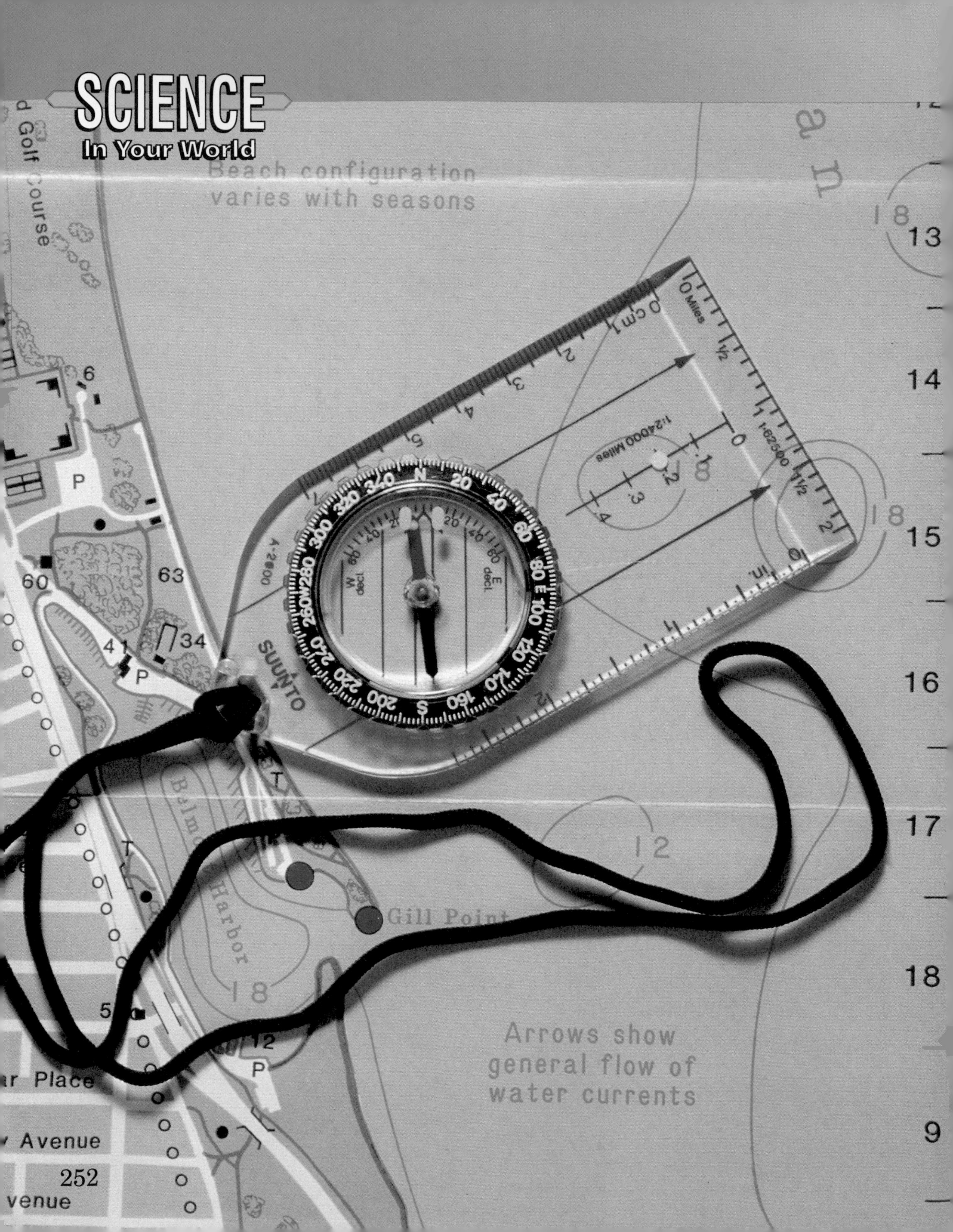
SCIENCE
In Your World
Beach configuration
varies with seasons
Golf Course
Arrows show
general flow of
water currents
Gill Point
Harbor
Place
Avenue
venue
SUUNTO
0 cm 1 2 3 4 5
0 Miles
1:24000 Miles
1:62500
13
14
15
16
17
18
9

CHAPTER 12

Magnetism and Electricity

Which way is north? A compass can tell you. The needle inside a compass is a little magnet that is being pulled by a larger magnet, Earth. Earth has a strong magnetic field. It will always pull the north-seeking end of a magnetic needle toward the north magnetic pole.

Have You Ever...

Joined Two Magnets?

Use chalk to label N or S on opposite ends of a craft magnet. Then use scissors to cut the magnet into two pieces. What can you do to prove that you have made two smaller magnets?

Try joining the two pieces again to form one magnet. Does it matter how the poles of the smaller magnets go together?

Magnetism

LESSON 1 GOALS
You will learn
- what a magnet is.
- how magnets affect other materials.
- what a magnetic field is.

Is there someone you know whom everyone likes? If so, you might say that person has a magnetic personality. People are drawn to that person by a force no one can see.

Magnets

A **magnet** is an object that is able to attract or pull some materials to it. Magnetic forces cause magnets to attract or repel other magnets. Magnets can also attract certain other objects.

Magnets come in many shapes and sizes. Think of all the ways you use magnets. Perhaps they hold notes on your refrigerator. Carpenters and mechanics use magnets to retrieve metal screws from places hands can't reach. Computer disks have magnetic materials that store images. There are hundreds of uses for magnets.

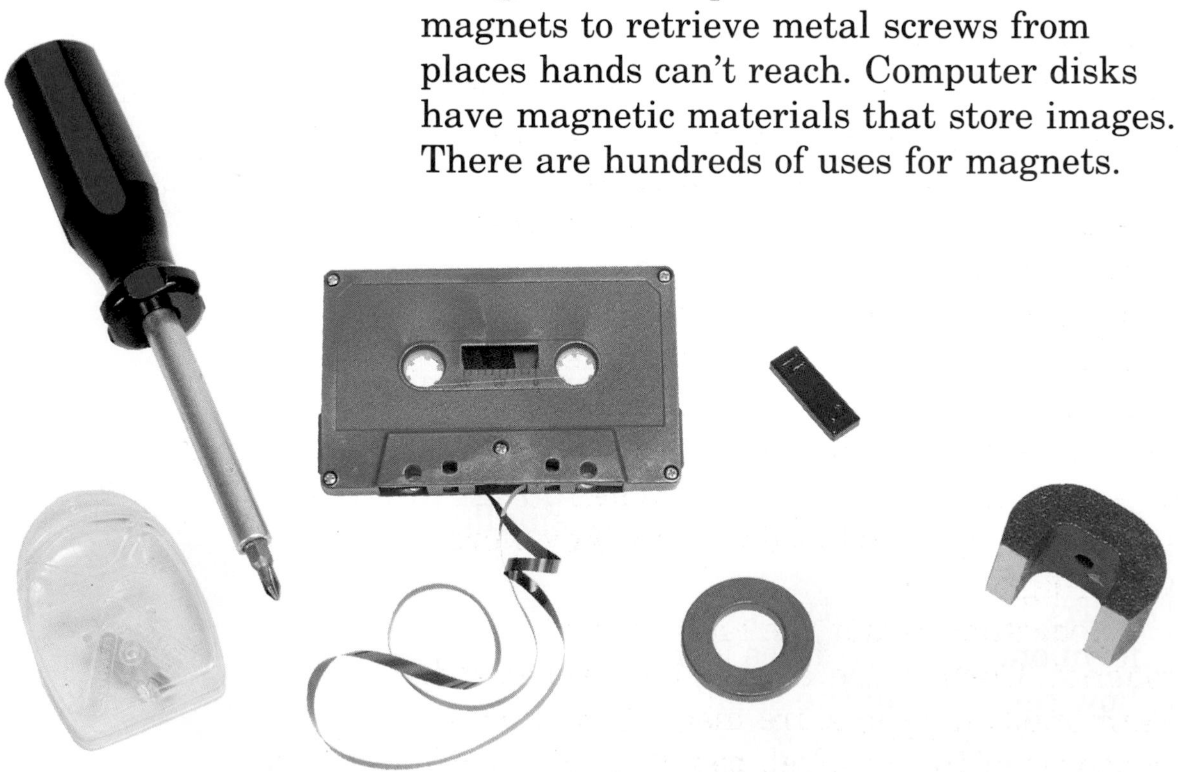

A variety of magnets

ACTIVITY

You Can...

Make a Magnetic Compass

Magnetize a needle by slowly stroking it in one direction with a strong magnet. Cut a piece of the bottom of a foam cup. Tape the needle to the middle of the foam and float it in a bowl of water. In which direction does the needle point? Turn the foam so the needle is in an east-west direction. What happens?

Where were magnets first discovered?

Magnets were first discovered in Magnesia, an ancient city in Greece. The people in Magnesia noticed that certain stones caused metals to stick to them. These stones were natural magnets, or **lodestones.**

In ancient times, sailors used lodestones to guide them at sea. The lodestones were floated on pieces of cork. They always moved themselves in a north-south direction. Because of this movement, the sailors could tell what direction they were traveling.

A lodestone is not the only kind of magnet. Certain metals can be magnetized. Permanent magnets are made from steel or mixtures of iron, nickel, cobalt, and other materials.

Would You Believe?

Tiny magnets that are as fine as powder are used to store images and sounds on tapes and computer disks.

Other materials, like copper, wood, glass, and rubber, can't be used to make magnets. Magnets don't attract them, either. You can test whether an object is magnetic or not by touching a magnet to it.

You can use a magnet to make any magnetic type of metal into a magnet. To magnetize an iron nail, rub the nail with the magnet. Move the magnet slowly in one direction only. Be sure to use the same end of the magnet each time. Heating or dropping the nail can cause it to lose its magnetism.

Like the lodestones ancient sailors used, compasses are really magnets. Inside a compass, a magnetic needle balances on a pin. One end of the needle points toward the north magnetic pole of Earth and the other end points toward the south magnetic pole of Earth. Because Earth is a giant magnet, it exerts weak magnetic forces on the compass needle and other magnets.

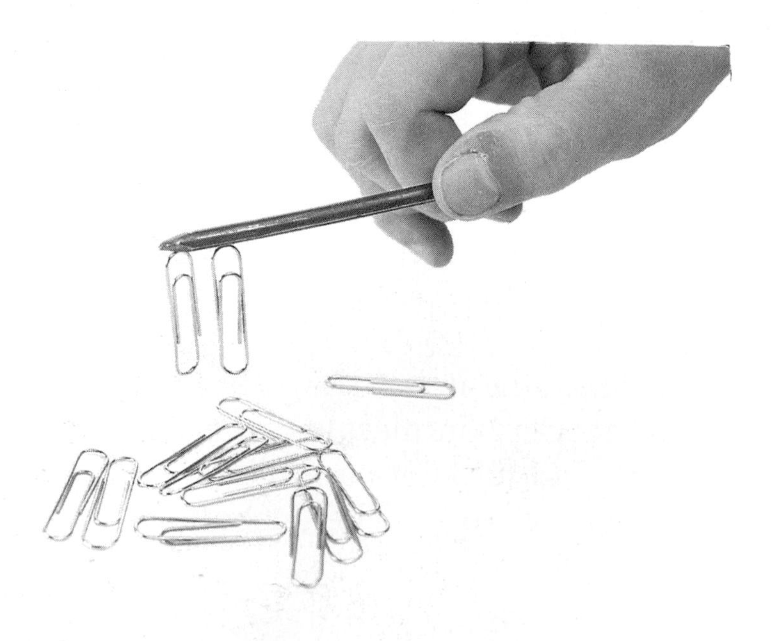

Some objects can become temporary magnets.

Many bar magnets have one end labeled "N" for north-seeking. The other end is labeled "S" for south-seeking. The ends of magnets are called **poles.** The magnetic force in a magnet is greatest at its poles. If you suspend a bar magnet on a string, the end labeled "N" will turn toward the Earth's north magnetic pole.

A north-seeking and a south-seeking pole, or two unlike poles, are attracted to each other. What happens when you place two north-seeking or two south-seeking poles (two like poles) near each other?

Magnetic Fields

What is a magnetic field?

The area around a magnet where the magnetic force acts is a **magnetic field.** Although you cannot see a magnetic field, you can see its effect. When iron filings are placed near a magnet, they form a pattern in the shape of the magnetic field.

You may have noticed that there are certain areas of a magnet where the magnetic force seems to be stronger. The pattern of the iron filings also shows that the magnetic force is stronger at the poles of the magnet.

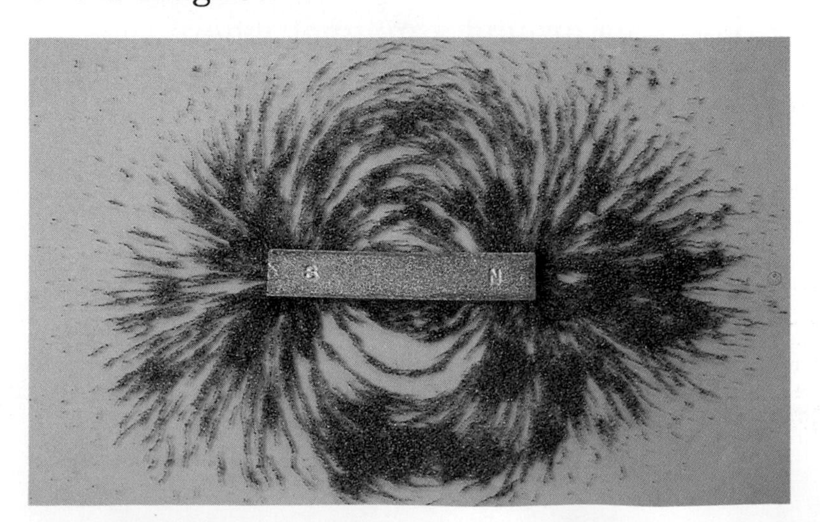

The pattern of iron filings shows the magnetic field.

In the 1500s, Francis Bacon, a famous Englishman, wrote about magnets. He said that the discovery of magnets and their usefulness in navigation was one of three discoveries that "have changed the whole face and state of things throughout the world." The other two were printing and gunpowder. Today we think nothing of using magnets. But there may be other uses for them that no one has thought of.

Lesson Summary

- A magnet is an object that attracts some materials.
- Magnetic forces control the behavior of magnets.
- The area around a magnet where the magnetic force acts is the magnetic field.

Lesson Review

1. Of what materials are most magnets made?
2. How do the poles of magnets affect one another?
★3. How are magnets useful to people?

Use Application Activity on page 371, 372.

How can you find the relative strength of magnets?

What you need

3 magnets (of different strengths)
meter stick
clear tape
compass
pencil and paper

What to do

1. Tape the compass on the 50 cm mark of the meter stick. Have the east and west sides of the compass point to the ends of the meter stick.
2. Rotate the meter stick so that the compass needle points directly north and south.
3. Put the first magnet on the 0 cm mark of the meter stick. Slowly move the magnet on the meter stick until the compass needle points toward the magnet.
4. Measure and record the distance from the magnet to the compass.
5. Repeat steps 3 and 4 using the other two magnets.

What did you learn?

1. How far was each magnet from the compass when the needle pointed east and west?
2. Which magnet was the strongest? The weakest?

Using what you learned

1. What might happen if you are using a compass and a strong magnet is nearby?
2. How could you determine if two magnets are stronger than one?
3. When might you use a weak magnet? A strong magnet?

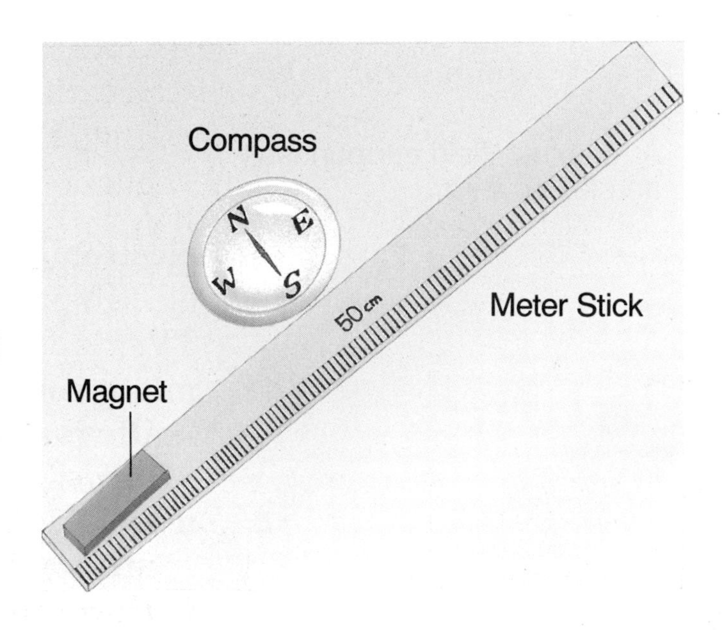

Magnets and Electricity

LESSON 2 GOALS
You will learn
- what an electromagnet is.
- how an electromagnet is made.
- how electromagnets are used.

Did you know that an electric current can cause a compass needle to move? The first time it was discovered that magnets and electricity affected each other was in 1820. A man from Denmark, Hans Christian Oersted, placed a compass under a wire that was carrying an electric current. He noticed that the compass needle moved, and he made the hypothesis that electric current affects magnets.

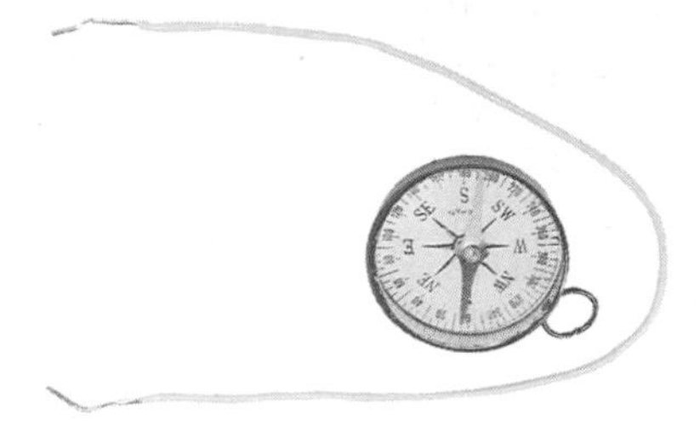

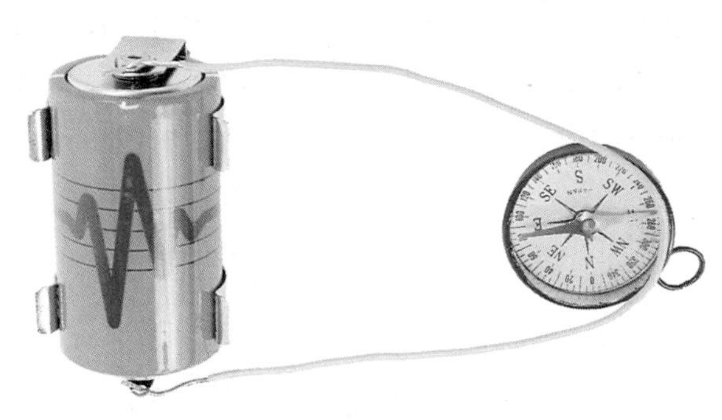

A magnetic field surrounds an electric wire.

You may have heard the word electromagnet (ih lek troh MAG nut) or electromagnetic when people talk about TVs, tape recorders, or other appliances. It might sound confusing to you, but it is really just the combination of two words, *electric* and *magnetic*. It was first used when Oersted made his discovery. An **electromagnet** is a temporary magnet that is made by using electricity.

What is an electromagnet?

You know that people use magnets to make other magnets. What Oersted showed us is that electricity can also be used to make temporary magnets called electromagnets. The movement of electrons along a wire produces a magnetic field around the wire. The magnetic field becomes stronger if the wire carrying electric current is wound into a coil. If you look inside an electric motor, you can see coils of copper wire wound around a metal rod. This is an electromagnet.

How can you make an electromagnet?

To make a simple electromagnet, you can wrap a wire around a nail. Then attach the ends of the wire to a battery to make an electric circuit. As current flows through the coil, a magnetic field is produced. The nail is magnetized and attracts certain metal objects. When the circuit is broken, the current stops. The nail is no longer magnetized.

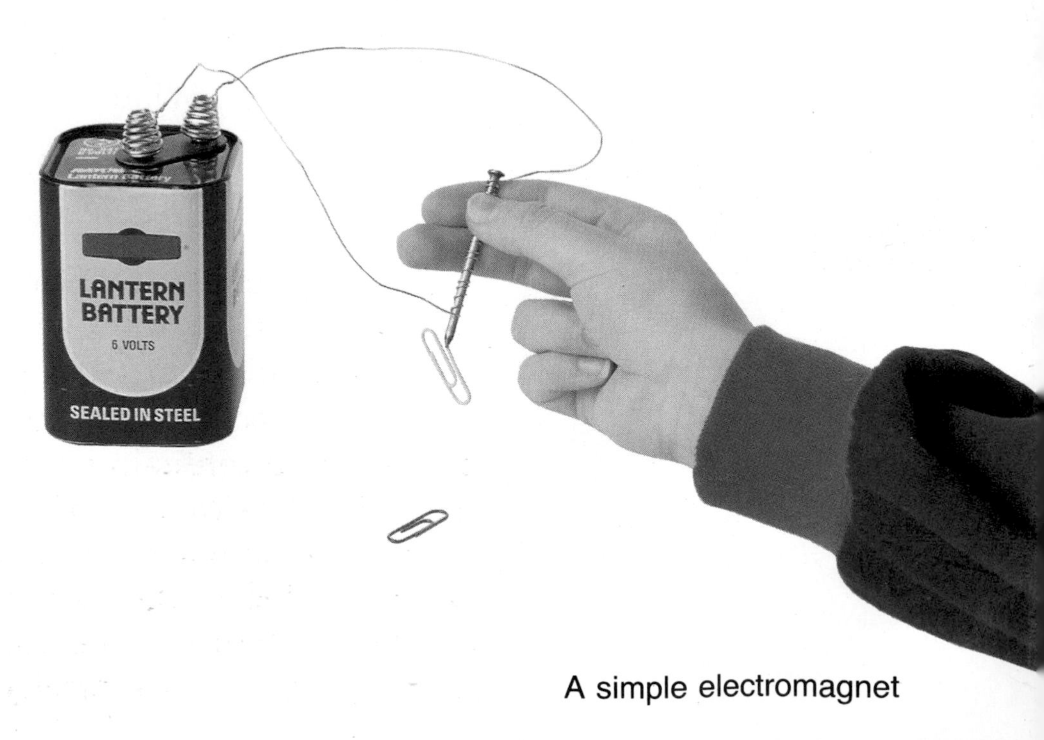

A simple electromagnet

Why would we want to use electromagnets? Electromagnets allow us to turn magnetic fields on and off. We can then control the magnetic energy. Scrapyard cranes, for example, use electromagnets. When an electric current flows through the coil of wire inside the plate on the crane, the plate becomes magnetized. The crane with its magnetized plate can then lift scraps of metal. The magnetism stops when the current is turned off.

Electromagnets are also used in doorbells, television and stereo speakers, washers and dryers, and other appliances with electric motors. Think of all the places where you find electric motors in use.

You might think you could live without electric energy. We all could do without certain appliances. But today many people's jobs depend on electric energy. Most people use electricity in their work in the form of lights, computers, typewriters, and power tools. What would happen if the electricity were turned off in your school?

Some clocks have electromagnets.

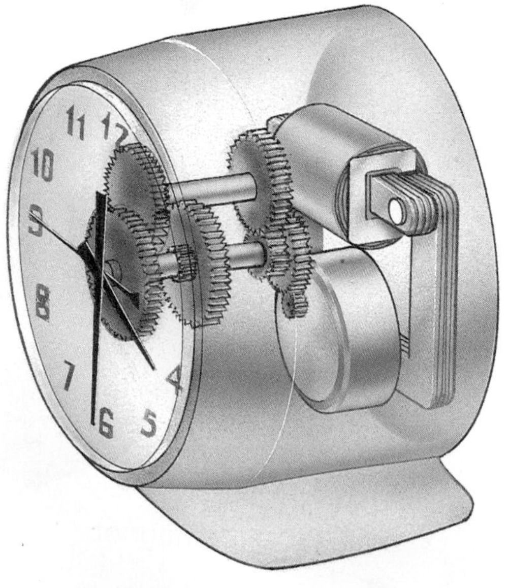

Repairing a circuit board

Electricians, repair people, sales people who sell electric-powered appliances, as well as scientists, engineers, and assemblers who make electric appliances are all directly affected by electromagnets.

Now think of all the things that need electricity to run. Electricity is a major energy source in homes, schools, factories, hospitals, and many other places. People use so much electricity every day that batteries alone couldn't make enough. Where else does electricity come from?

Over the years people have built different types of power plants, which are producers of large amounts of electricity. Power plants use another effect of electromagnetic energy to create electricity. The movement of a magnet in a coil of wire causes an electric current to flow through the wire. A **generator** (JEN uh rayt ur) is a machine that produces an electric current when a coil of wire turns within a magnetic field.

What is a generator?

SCIENCE AND . . .
Reading

Which is an OPINION from the chapter?

A. We all could do without certain appliances.
B. All magnets have two poles.
C. Some people are concerned about nuclear power.
D. If a circuit breaks, current stops.

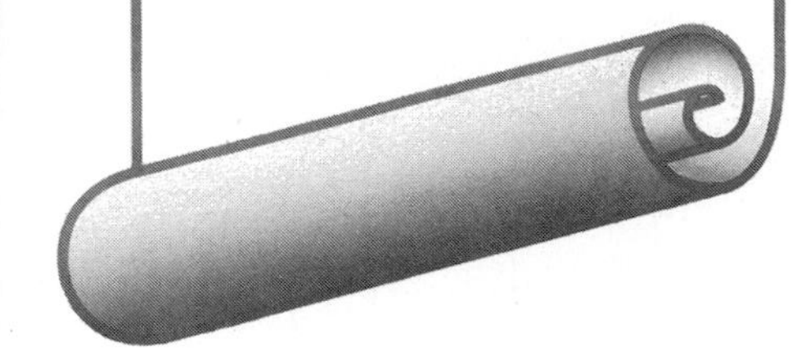

How are turbines used within generators to produce current electricity?

In power plants, steam turbines rapidly spin the large coils of wire within the magnetic fields of huge electric generators. The amount of current that is generated depends on the strength of the magnetic field of a generator's magnets. It also depends on the number of turns of wire in the coil and the speed at which the coil is turned.

Turbines provide the power for the generators to produce electricity. A turbine has blades attached to an axle. The power to turn the turbine blades comes from the pressure of steam. The steam rushes through the turbine, causing the blades to turn rapidly. The turbine is connected to the coils of a generator. The circular motion of the blades and axle causes the generator's huge coils of wire to spin. The coils spin in the generator's strong magnetic field. This produces electric energy in the coils. Power lines connected to the generator carry current electricity from the power plant to towns and cities.

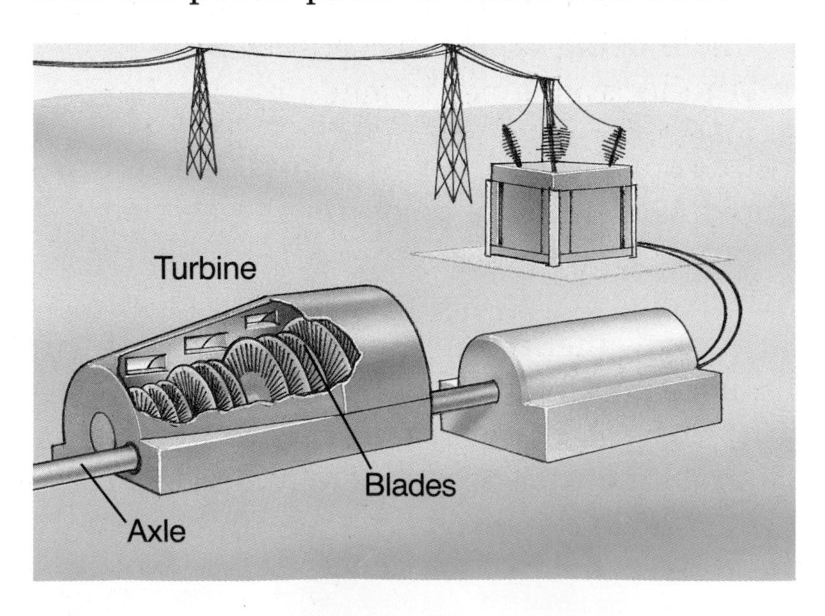

Parts of a generator

Wind can be used to produce electricity.

In most cases, the steam needed to run a steam turbine is produced by burning such fuels as coal, oil, or natural gas. In nuclear power plants, energy from uranium atoms is released and used to heat water and produce steam. No matter which fuel is used to produce electricity, there are possible problems regarding pollution of the environment. Perhaps you will be the one to discover the fuel of the future.

Lesson Summary

- An electromagnet is a temporary magnet made using an electric current.
- An electromagnet is made by passing electric current through wire that is wrapped around a metal rod.
- Electromagnets are used in power plants and in most electric devices.

Lesson Review

1. How does an electromagnet work?
2. What fuels are used to produce electricity?

★3. How are electromagnets useful to us?

How can you make an electromagnet?

What you need

2 fresh batteries
2 pieces of bell wire (1 meter and 15 cm)
battery holder
paper clips
switch
nail
pencil and paper

What to do

1. Wrap the 1-meter wire around the nail 15 turns as shown.
2. Connect the battery, switch, and nail as shown.
3. Record how many paper clips you can pick up with the electromagnet.
4. Repeat steps 2 and 3 using two batteries.
5. Repeat using 30 turns of the wire.

What did you learn?

1. How many paper clips did you pick up each time?
2. When was your electromagnet strongest?

Using what you learned

1. When does the nail lose its magnetism?
2. How is an electromagnet different from a permanent magnet?
3. Does the number of wire turns or number of batteries affect strength more?

Paper clips held		
	15 turns	30 turns
1 battery		
2 batteries		

I WANT TO KNOW ABOUT...

Comparing and Contrasting

This metal is as cold as ice!

One way to describe something is to tell how it is like something else. Another way is to tell how it is different. A comparison shows how things are alike. A contrast shows how they are different.

We compare and contrast things all the time. It might be colder today than yesterday. Today's lunch might be better than yesterday's lunch.

You might use comparison and contrast in science to describe how magnets affect different objects. Collect a group of at least ten different small things. Then, take a magnet and touch each one. Sort the objects into two groups depending on whether they are attracted to the magnet or not. You may then compare how strongly each object is attracted to the magnet.

Finally, choose one of the objects and describe it to a friend or the class without telling what it is. First, tell how it is like the other things you tested with a magnet and then how it is different. Then, you can describe its color, size, and function. See if other people can guess what it is.

Language Arts

CHAPTER REVIEW 12

Summary

Lesson 1

- A magnet is an object that attracts some materials.
- Magnetic forces control the behavior of magnets.
- The area around a magnet where the magnetic force acts is the magnetic field.

Lesson 2

- An electromagnet is a kind of temporary magnet.
- An electromagnet is made of wire wrapped around a metal rod.
- Electromagnets are used in most electric devices.

Science Words

Fill in the blank with the correct word or words from the list.

magnet
lodestones
poles
magnetic field
electromagnet
generator
turbines

1. Stones that are natural magnets are called ___.
2. The area around a magnet where the magnetic force acts is a(n) ___.
3. An object that is able to attract or pull some objects to it is a(n) ___.
4. The ends of magnets are called ___.
5. A machine that produces an electric current when a coil of wire turns within an electric field is a(n) ___.
6. A temporary magnet made by using electricity is a(n) ___.
7. The power needed for generators to produce electricity is provided by ___.

Questions

Recalling Ideas

Correctly complete each of the following sentences.

1. Magnets can't be made from
   (a) copper. (c) wood.
   (b) rubber. (d) all of these
2. Permanent magnets are made from
   (a) plastic. (c) steel.
   (b) glass. (d) wood.
3. You can determine whether an object is magnetic or not by
   (a) heating it.
   (b) touching a magnet to it.
   (c) dropping it.
   (d) touching it to glass.
4. A major energy source in homes and schools is
   (a) electricity.
   (b) magnets.
   (c) electromagnets.
   (d) batteries.

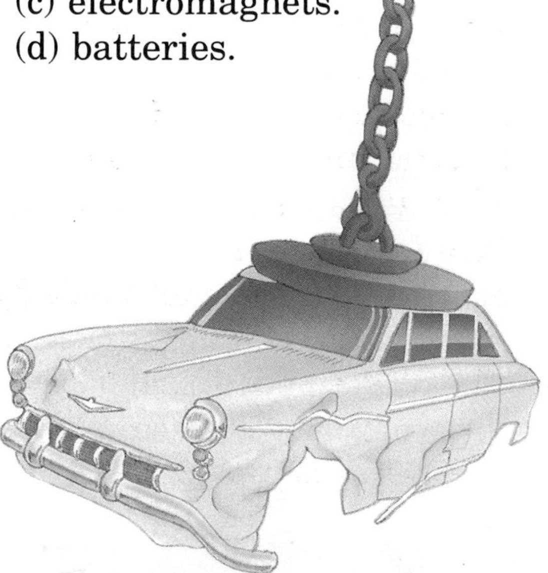

5. The steam needed to run a steam turbine is probably produced by
   (a) coal. (c) natural gas.
   (b) oil. (d) all of these

Understanding Ideas

Answer the following questions using complete sentences.

1. What hypothesis was made by Hans Christian Oersted?
2. What is the origin of the word *electromagnetic?*
3. When are electromagnets useful?
4. What are some uses of electromagnets?
5. What determines the amount of current generated by generators in power plants?

Thinking Critically

Think about what you have learned in this chapter. Answer the following questions using complete sentences.

1. Why do two magnets seem to push away from each other when you try to put them together?
2. How can you increase the strength of an electromagnet without changing the current in the wire?

UNIT REVIEW 3

Checking for Understanding

Write a short answer for each question or statement.

1. Describe the two things that can happen to light as it strikes an object.
2. How is a rainbow formed?
3. Why does a leaf appear to be green?
4. Name and describe three properties of sound.
5. What causes static electricity?
6. What is current, and how can it be controlled in a circuit?
7. Explain how magnetic forces control the behavior of magnets.
8. Compare a permanent magnet and an electromagnet.
9. What are some uses of electromagnets?
10. What is a generator?
11. What is light?
12. What happens to light reflected from a smooth surface? A rough surface?
13. When is light refracted?
14. Give one example of each: transparent, translucent, and opaque matter.
15. What is the visible spectrum?
16. Give two examples of how color is used in everyday living.
17. How does a prism separate white light?
18. How do you hear sound?
19. What kinds of surfaces reflect sound?
20. Through what kind of matter does sound travel fastest?

Recalling Activities

Write a short paragraph for each question or statement.

1. How can you bend light?
2. What do sound vibrations look like?
3. What is static electricity?
4. How can you make the bulb light?
5. How can you find the relative strength of magnets?
6. How can you make an electromagnet?

Project Ideas

1. Display some insulators and conductors, and demonstrate and report on their different uses.
2. Compare the colors that objects appear to be in different colors of light.
3. Set up a demonstration to compare the sound carrying abilities of different materials.
4. Build and demonstrate a simple generator. Show how the current can be increased or decreased and how it can be used.

Books to Read

Sound Experiments by Ray Broekel, Children's Press: Chicago, 1983.
Fun experiments to help you learn about sound.

Discovering Electricity by Neil Ardley, Franklin Watts: Danbury, CT, 1984.
This book describes investigations in electricity.

Exploring With Lasers by Brent Filsin, Julian Messner: New York, 1984.
Read this book to learn more about lasers.

Whether I'm dirty or whether I'm not,
Whether the water is cold or hot,
Whether I like or whether I don't,
Whether I will or whether I won't,
"Have you washed your hands, and washed
your face?"
I seem to *live* in the washing-place.

from "Washing"
John Drinkwater

RUNNING
ROUND ROCK
ROADRUNNERS

Taking Care of Yourself

Some athletes barely make it to the finish line. Others seem to have energy to spare. But eventually all athletes will need to rest. The amount of time a body can exercise before needing rest is called *endurance*. Amazingly, the more you exercise, the greater your endurance will be.

Have You Ever...

Tested Your Endurance?

If you think holding a book is an easy task for your body to endure, try this.

Hold a heavy book in one hand while extending your arm. Can you hold this position for two minutes? What do you need to do to accomplish this task?

Exercise for Physical and Mental Health

LESSON 1 GOALS
You will learn
- how exercise improves physical health.
- how exercise improves mental health.

Having a healthy body doesn't just happen. You have to know how to take care of your body and then do something about keeping your body healthy. One part of having a healthy body is getting regular exercise. **Exercise** is any activity that uses the muscles of the body.

If you were to ask your friends what they do for exercise, what would they say? When Lisa asked her friends to tell her what they do for exercise, she found they had many different activities. For example, Rosa practices three days a week with her soccer team at school. Matthew rides 20 minutes on his bicycle to and from school every day. Seth and Ann take dancing lessons twice a week and practice between lessons.

Regular exercise helps keep your body healthy.

Lisa decided that her friends were exercising regularly. Some of them were exercising more than others, but each was spending a regular amount of time using his or her muscles in some activity. What kinds of exercises do you regularly do to improve your physical and mental health? In this lesson, you will learn how exercise helps you keep your body physically and mentally fit.

You exercise while playing some games.

You may have a toy or game that works like a machine. For your machine to work the way it should, all of its parts must work well. When one part doesn't work at its best, the other parts can't work as well. Your body is like a machine. For you to work at your best, you have to make sure each part of your body works at its best.

Exercise can be fun.

Exercise is important for your overall physical health. You can choose activities that will improve your circulation, strengthen your muscles, and improve your flexibility. The best part about exercise is that you can have fun while you are taking care of yourself.

Physical Health

What is circulation?

How can exercise improve your circulation? Circulation is the movement of blood flowing through your body. Your heart is the most important organ involved in circulation. Your heart is a muscle that pumps blood all through your body. Your heart must be strong so that blood can reach your skin, arms, legs, brain, and other body parts.

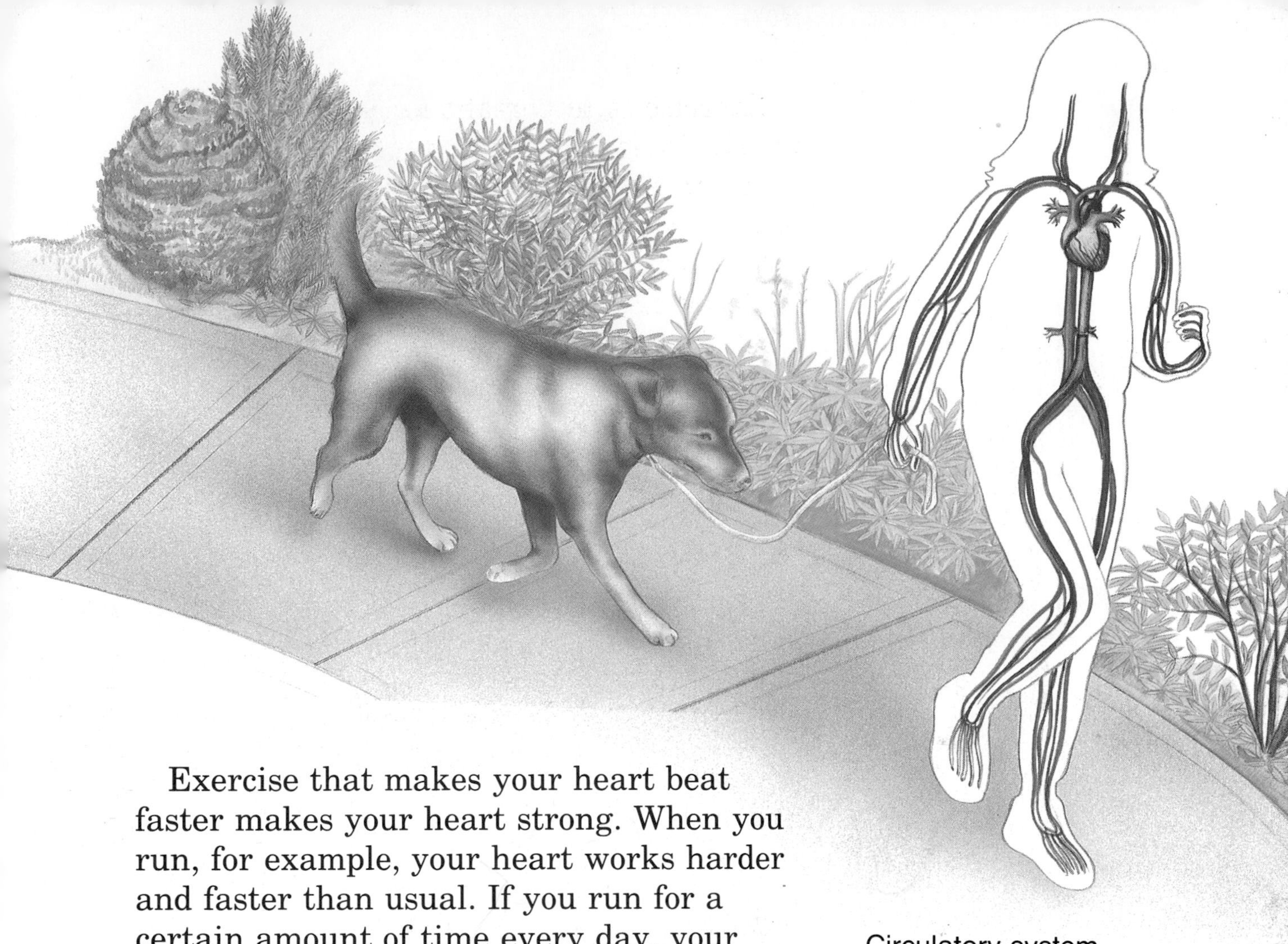

Circulatory system

Exercise that makes your heart beat faster makes your heart strong. When you run, for example, your heart works harder and faster than usual. If you run for a certain amount of time every day, your heart gets used to doing hard work. It doesn't have to pump as many times because it pumps more blood with every beat. As a result, running becomes easier for your heart and your circulation improves.

Exercise also makes your skeletal muscles stronger and firmer. Over a period of time, your body becomes used to exercise. With regular exercise, you may notice that you can work and play harder for longer periods of time without getting tired. You probably can move and stretch more easily without hurting yourself. Your joints move more freely.

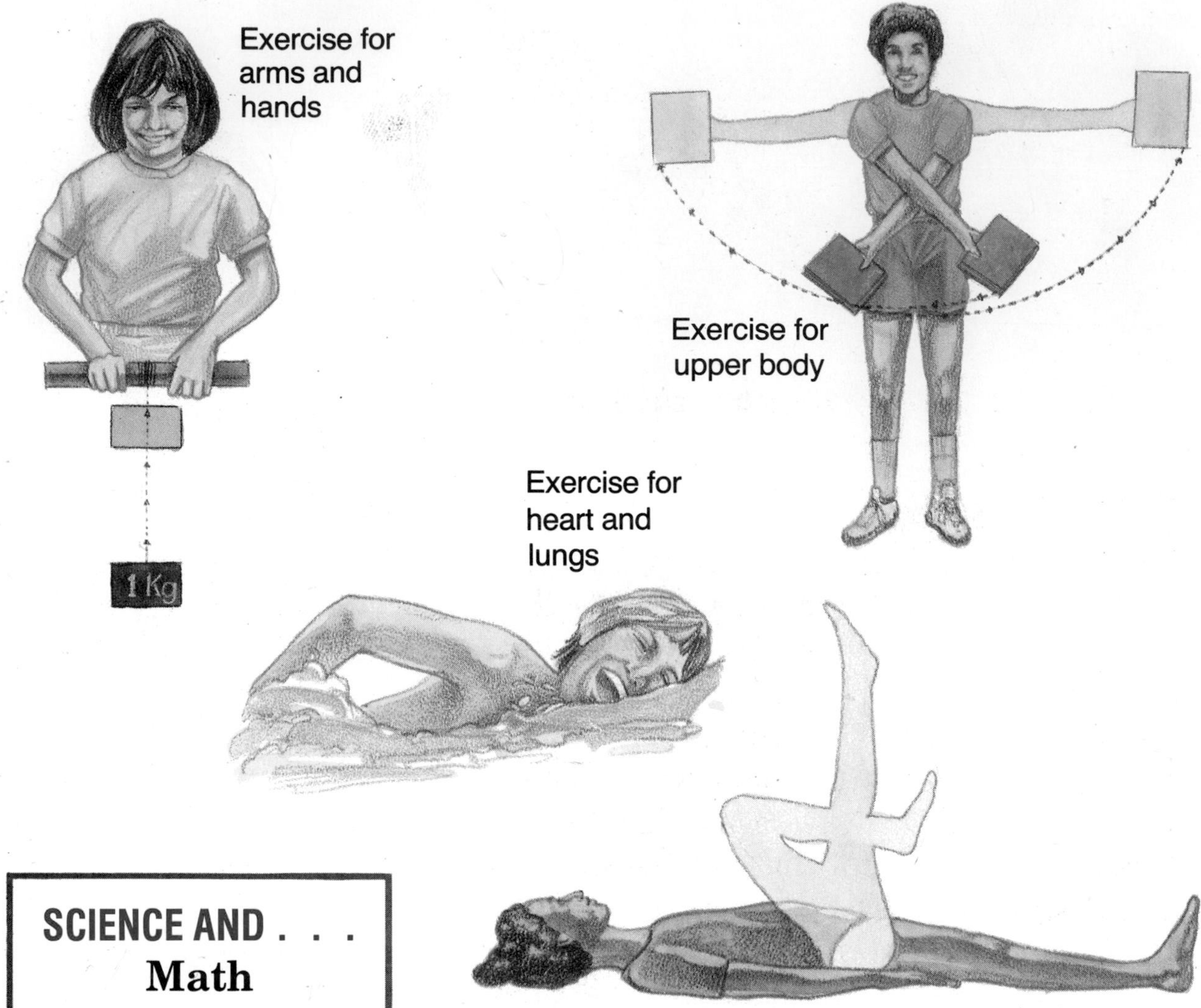

Exercise for stomach and waist

SCIENCE AND . . .
Math

Kim swims more than Kay. Kay swims less than Ian. Max swims more than Kim. What is most reasonable?

A. Max swims more than Kay.
B. Ian and Kim swim the same amount.
C. Max swims less than Ian.

You have many different muscles in your body that need to be exercised to keep them strong. You can exercise the different muscles in your body in many ways. Running, swimming, walking, dancing, and bicycling are exercises that make the heart stronger. Pushing, lifting, and pulling are types of exercises that strengthen your skeletal muscles. The pictures on this and the next page show some exercises for strengthening the whole body.

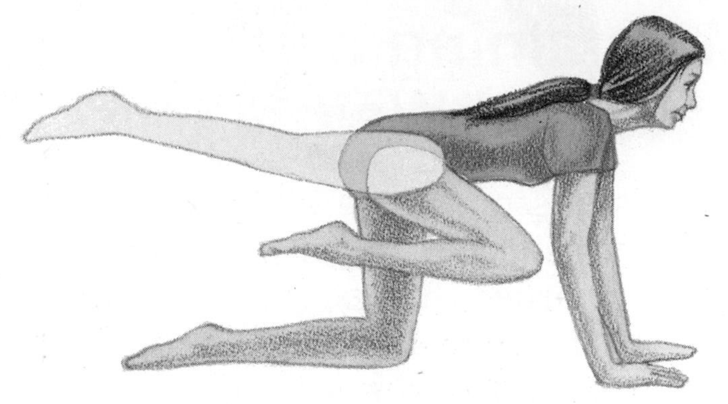
Exercise for lower back and thighs

Exercise for hips, legs, feet

Mental Health

When you exercise, you strengthen your whole body. When your body is in good condition, you feel better in many ways. You reduce the amount of stress that you feel with certain tasks. You can think more clearly and show your emotions in more healthful ways. Perhaps best of all, you think and feel better about yourself because you have done something good for YOU!

Lesson Summary

- Exercise makes your heart and skeletal muscles healthier.
- Exercise improves the way you think and feel about yourself.

Lesson Review

1. What kind of exercises make your heart strong?
2. How does exercise improve your skeletal muscles?
★3. How is your mental health improved by exercising?

Maintaining a Healthful Weight

LESSON 2 GOALS
You will learn
- what Calories are and how unused Calories affect the body.
- how you can maintain a healthful weight.

Lisa's classmates are interested in keeping their bodies healthy. They have already learned that exercise helps them improve their physical and mental health. They want to know other ways that exercise helps to keep their bodies healthy. How much exercise do they need to stay healthy?

Before we can answer this question, we need to think about what a healthy body needs. When you exercise, or do any activity, you use energy. How much energy you use depends on how active you are.

You use energy throughout the day.

The energy you use comes from the food you eat. A **Calorie** (KAL uh ree) is a measure of the amount of energy in foods. Foods like cakes, pies, and other desserts are high in Calories, while foods such as vegetables are low in Calories. You can get an idea of the number of Calories in some of the foods you eat by looking at Table 1 "Calories in Food." To see how many Calories you use per hour doing certain activities, look at Table 2 "Calories Used Per Hour."

Table 1 Calories in Food

Food	Calories
1 apple	80
1 slice cake	180
3 pancakes with syrup	350
3 slices pizza and 1 cola drink	580
1 glass milk	165
1 pork chop	260
1 yam	155
1 tomato	30

Table 2 Calories Used Per Hour

Activity	Calories Used
eating	84
reading	72
watching TV	80
swimming	300
walking	216
dancing	330
sleeping	60
bicycling (fast)	250

Suppose during one day you ate all the foods listed in Table 1. What would be the total number of Calories you ate? Suppose during the same day you used energy to eat three meals, watch TV for three hours, and sleep for eight hours. How many Calories would your body use for all of these activities?

Why does the body need energy?

Your body needs the energy from Calories in order to do all the activities you do each day. But it also needs some energy for growth. How do you know your body is using Calories for your growth? Have you noticed that the sleeves on your shirt you wore last year are too short? You may be a few inches taller than you were last year. As you grow, your body gains weight.

What happens to the Calories that your body doesn't need to grow or to do your daily activities? These Calories are stored by the body as fat. Fat is used by your body in many ways, such as to protect your organs and to keep you warm. So, fat is not all bad. But, when your body stores too much fat, your body gains too much weight. Too much weight makes your heart and body work harder than they should.

You use Calories to grow.

When Lisa and her classmates exercise regularly, their bodies use more of the Calories they eat. Exercise helps them control their body weight. Their bodies don't have to store too many unused Calories. When they eat the amount of food their bodies need for growth and daily activities, they are developing good eating habits, as well as controlling their body weight. Exercise is a good way to have a balance between the Calories your body needs for growth and daily activities and the Calories you eat.

Balance your meals and activities.

A healthful lifestyle

Lesson Summary

- The body stores unused Calories from food as fat.
- You can maintain a healthful weight by balancing the amount of food you eat with your level of activity. Consider that you need extra Calories while growing.

Lesson Review

1. What is a Calorie?
2. How is your weight affected by your body's growth?

★3. Why is too much stored body fat a problem?

How many Calories do you use in one week?

What you need

Calorie table
pencil and paper

What to do

1. Make a chart like the one shown.
2. Keep track of how much time you spend doing each activity every day for a week.
3. Add up and record the total time you spend on each activity for the week.
4. Use the Calorie table on page 283 of your textbook. Figure out how many Calories you used to do each activity.

What did you learn?

1. How much total time did you spend eating? Watching TV?
2. How many Calories did you use walking? Sleeping?

Using what you learned

1. Why is it important to balance the Calories your body uses with the Calories you eat?
2. Why is regular exercise important?

Activity	Time spent per week							Total time
	Mon.	Tues.	Wed.	Thurs.	Fri.	Sat.	Sun.	
Eating								
Sleeping								
Walking								
Reading								
Watching TV								

Effects of Sleep

LESSON 3 GOALS
You will learn
- how sleep affects your body.
- why sleep is important for a healthy body and mind.

Another important part of taking care of your body is letting your body get enough rest. Why is it important for your body to rest?

After a machine works for a long time, the motor tends to get very hot. It needs time for the parts to cool. It may need daily repair. Your body is similar to a machine in this way, too.

After a full day, your body needs sleep. **Sleep** is a state of restfulness. When you sleep, your muscles relax. Your heart beats fewer times per minute. You also take fewer breaths per minute as you sleep. Your brain gets rest from thinking and remembering, too.

Sleep—a state of restfulness

Everyone needs sleep, but in different amounts. Babies in the first few weeks after birth may sleep as many as 16 hours a day. Most adults need about 7 or 8 hours of sleep a day.

Most young people your age need 10 to 12 hours of sleep each night. There are several important reasons for getting enough sleep at your age. You need sleep for your body to rest after all the activities you do during the day. Your muscles become better prepared for the next day's activities. Also, you are growing rapidly. While you are resting, your body still is active. Worn-out body cells are replaced. Chemicals, called **growth hormones,** are released into your circulatory system. These hormones cause your body to grow. You could say you are growing as you rest and sleep!

Why does your body need sleep?

ACTIVITY

You Can...

Become a Sleep Scientist

Some scientists study sleeping behaviors. You can collect data on sleep using your own family. Make a chart to collect data. What interesting observations did you make? Ask some students in another grade to use the same chart. Do some people who are the same age need different amounts of sleep?

Sleep gives you peace of mind. Sleep helps you feel refreshed. Being refreshed means you should be able to think clearly and concentrate on your work.

Lack of sleep affects your ability to concentrate.

Sometimes you may need more rest than at other times. When you are ill, your body needs extra energy to heal itself. Even when you feel better, your body may still need more rest than usual to let it completely heal.

You need more sleep when you're ill.

You can see that sleep is not time lost from your day. With sleep, your body has a better chance of working the way it should. You feel better, happier, and healthier.

Lesson Summary

- During sleep, your muscles relax, your heart beats fewer times per minute, and your brain gets rest from thinking.
- Sleep allows your body time to rest. Worn-out body cells are replaced and growth hormones are released into your circulatory system, causing you to grow. Sleep helps you to be able to think clearly and concentrate.

Lesson Review

1. What is sleep?
2. How much sleep do most children your age need?

★3. Why may you need more sleep than usual while ill?

Would You Believe?

Most people have four to five dreams every night, lasting for a total of about two hours.

Personal Cleanliness

LESSON 4 GOALS
You will learn
- that cleanliness is important.
- what you can do to have a well-groomed appearance.

Lisa and her classmates have learned that healthy bodies don't just happen. They know that exercising, maintaining their body weight, and getting enough sleep are all important for taking care of themselves.

Personal cleanliness is also important for having good physical health. **Cleanliness** is the habit of being well-groomed. To remain healthy and well-groomed, you need to take care of your skin, nails, hair, and teeth. In this lesson, you will learn some actions to do to maintain a well-groomed appearance.

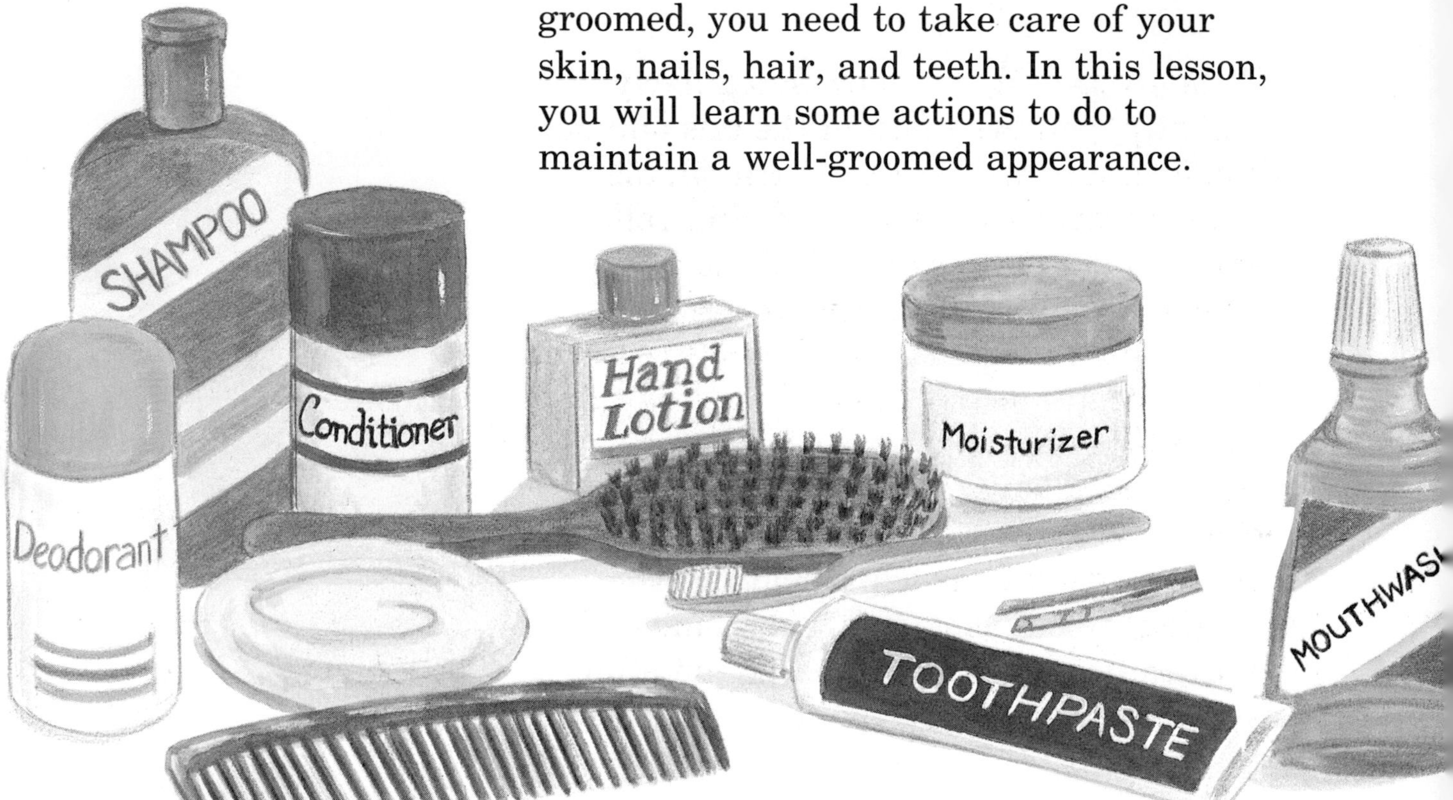

Skin Care

The outer layer of your skin is made of dead cells that are constantly being replaced by new cells. It is also a place where body oils, dirt, and perspiration collect.

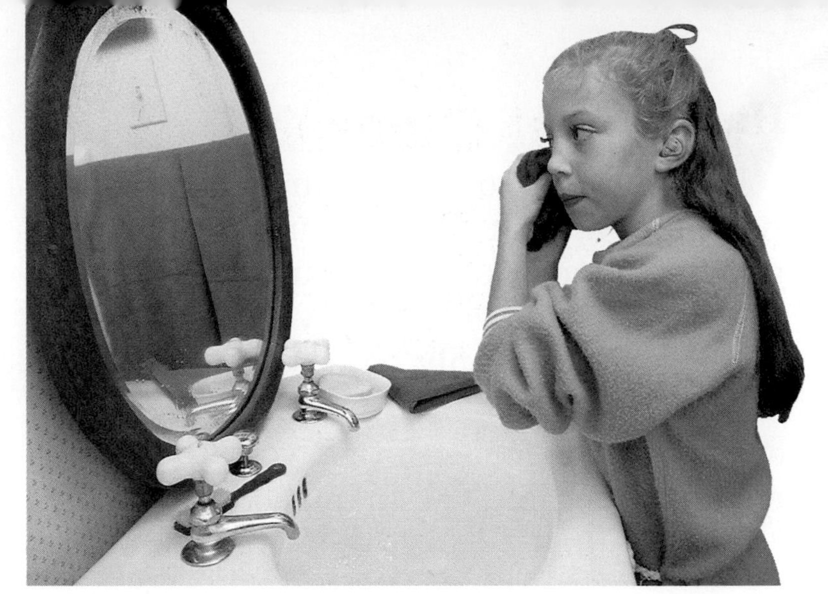

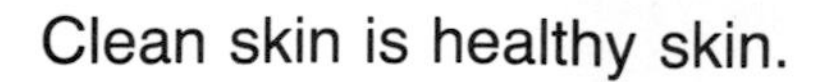
Clean skin is healthy skin.

When you bathe or shower regularly with soap, you break down the oils on your skin. You also wash away extra oil, dirt, germs, perspiration, and dead skin cells. Bathing or showering regularly is important for clean, healthy-looking skin.

The skin on your face and hands requires special attention. You should wash your face at least twice a day with soap and water, and wash your hands more often. Scientists have learned that hands spread more cold germs than coughing or sneezing. When you wash your hands before eating or preparing food and after using the bathroom, you help to remove the germs.

How are cold germs most often spread?

Nail Care

You should clean your fingernails and toenails daily. A nailbrush or nail file should be used to clean under the nails. Keep fingernails and toenails trimmed neatly and use a nail file to smooth away rough edges.

Hair Care

Use shampoo to wash your hair at least two times a week. When you shampoo, wet your hair and apply the shampoo. Then, rub it into your scalp with your fingertips and rinse well. Apply more shampoo, rub it in, and rinse well again. Be sure to rinse out all the shampoo so that your scalp will not itch from leftover shampoo, and your hair will shine.

Good hair care

Teeth

When you eat, particles of food may stick to your teeth or may be caught between your teeth. These particles may form plaque (PLAK). **Plaque** is a sticky material that forms on teeth and is harmful to dental health. Brushing and flossing are two ways to remove and prevent plaque from forming on your teeth. Brush your teeth at least two times a day.

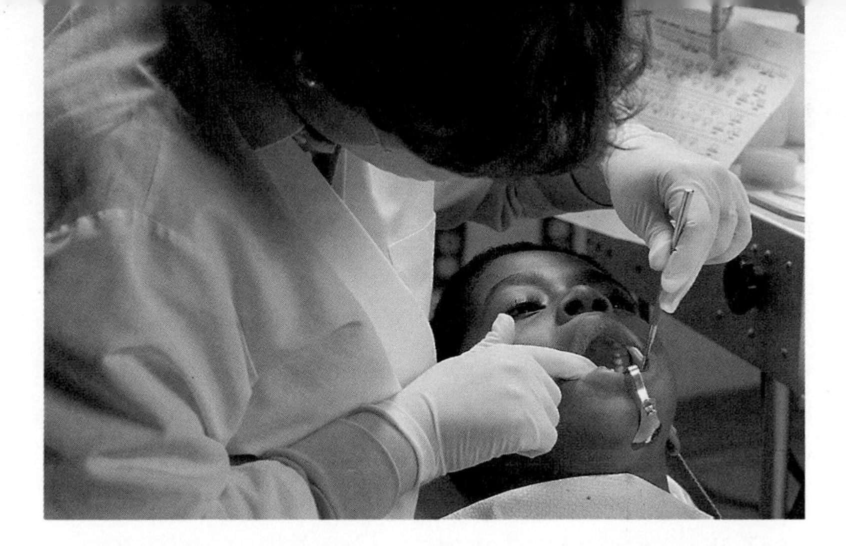

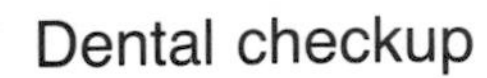

Dental checkup

Dental checkups and fluoride treatments are extra steps you can take to avoid problems. Taking care of your teeth is important for a well-groomed appearance.

When you meet someone for the very first time, you probably won't know much about him or her. But you know right away if that person cares about how he or she looks. When people meet you for the first time, will they know that you care about your appearance? Taking care of your appearance is important in maintaining a healthy body.

Lesson Summary

- Personal cleanliness promotes physical health.
- A well-groomed person takes care of skin, nails, hair, and teeth.

Lesson Review

1. Why is skin care important?
2. Why must you carefully rinse your hair after washing?
★3. What are some ways that personal cleanliness promotes physical health?

Why wash your hands?

What you need

2 pieces of bread
2 plastic bags (zip lock)
soap
water
paper towels
pencil and paper

What to do

1. Wet the fingertips of one hand. Rub them over dusty places in the classroom.
2. Fold a piece of bread in half. Rub your fingertips inside the folded bread. This is like putting your dirty fingers inside of your mouth.
3. Put the folded bread into a plastic bag. Close the bag and put it in a warm, dark place.
4. Wash your hands with soap and water. Dry them with the paper towels.
5. Rub your clean fingers inside a second piece of folded bread. Repeat step 3.
6. Observe the bread inside each bag after 3 and 5 days. Do not open the bags.

What did you learn?

1. What was in the bags after 3 days? After 5 days?
2. How does dirt get on your hands?

Using what you learned

1. Why should you wash your hands several times a day?
2. Why should you not put fingers in your mouth?
3. Why should you not chew on your pencil?

I WANT TO KNOW ABOUT...

Dental Hygienists

Charles Howard is a dental hygienist. He helps the dentist treat and care for the mouths, teeth, and gums of patients.

Charles sees several patients each day. He likes meeting different people. But, what he likes best about his profession is talking with children and showing them how to care for their teeth. Charles knows all about proper brushing, good nutrition, and the importance of regular checkups.

After he makes a patient comfortable in a reclining chair, Charles takes X rays of the teeth and gums. Next, he uses hand tools and machines to carefully remove stains and deposits from the patient's teeth. After the teeth are clean, he sometimes applies fluoride to the teeth to help prevent diseases of the gums and tooth decay.

Before seeing his next patient, Charles organizes the X rays and discusses the patient with the dentist.

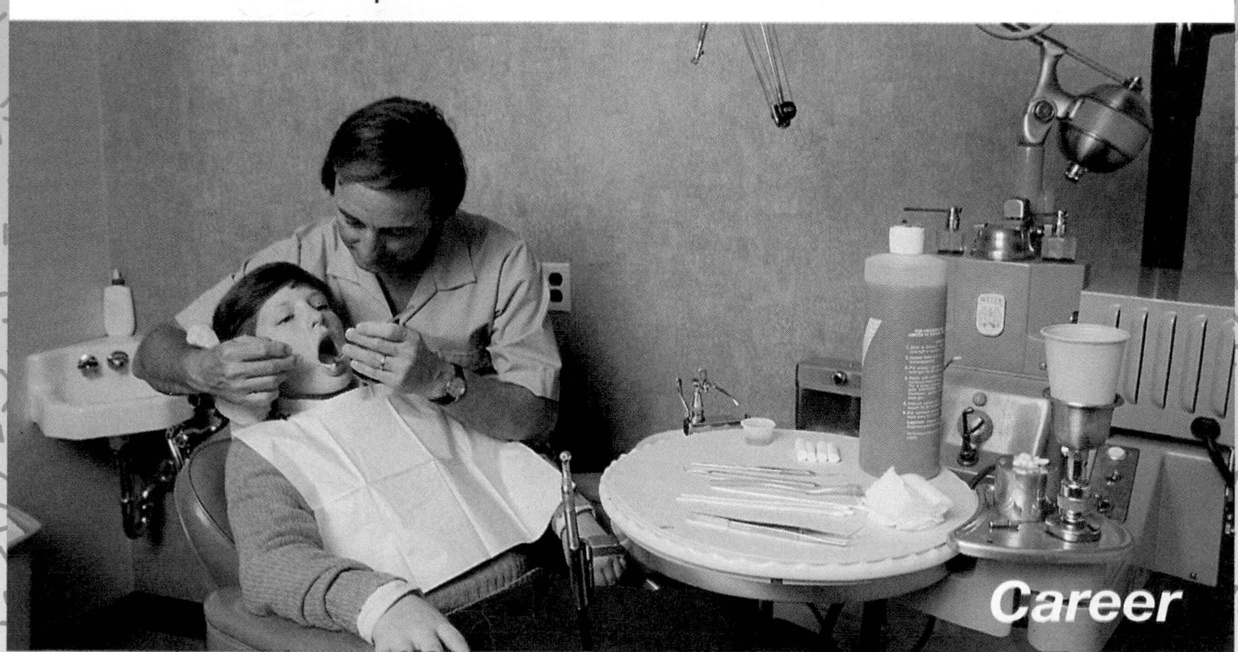

CHAPTER REVIEW 13

Summary

Lesson 1
- Exercise makes your heart and skeletal muscles healthier.
- Exercise improves the way you think and feel about yourself.

Lesson 2
- The body stores unused Calories from food as fat.
- You can maintain a healthful weight by balancing the amount of food you eat with your level of activity. Consider that you need extra Calories while growing.

Lesson 3
- During sleep, your muscles relax, your heart beats fewer times per minute, and your brain gets rest.
- Sleep allows your body to rest. Worn-out body cells are replaced and growth occurs. Sleep enables you to think clearly and concentrate.

Lesson 4
- Personal cleanliness promotes physical health.
- A well-groomed person takes care of skin, hair, nails, and teeth.

Science Words

Fill in the blank with the correct word or words from the list.

exercise	**sleep**	**plaque**
Calorie	**cleanliness**	**growth hormones**

1. The habit of being well-groomed is ____.
2. An ____ is a measure of energy in food.
3. Any activity that uses the muscles of the body is ____.
4. ____ is a state of restfulness.
5. A sticky material that forms on teeth and is harmful to dental health is called ____.

Questions

Recalling Ideas

Correctly complete each of the following sentences.

1. Exercises that make the heart stronger include
   (a) bicycling.
   (b) swimming.
   (c) running.
   (d) all of these
2. Extra Calories are stored as
   (a) muscles. (c) fat.
   (b) cartilage. (d) bone.
3. You can help control your body weight with
   (a) sleep. (c) growth.
   (b) exercise. (d) all of these.
4. The number of hours of sleep an average ten-year-old needs daily is about
   (a) 7 or 8. (c) 16.
   (b) 10 to 12. (d) 18.
5. To remain healthy and well-groomed, you need to take care of your
   (a) skin. (c) teeth.
   (b) hair. (d) all of these.

Understanding Ideas

Answer the following questions using complete sentences.

1. What changes might you notice about your body after it becomes used to exercise?

2. How can regular exercise improve your mental health?
3. How does too much weight affect your body?
4. How does sleep affect your body?
5. What are two ways to remove plaque and prevent it from forming on your teeth?

Thinking Critically

Think about what you have learned in this chapter. Answer the following questions using complete sentences.

1. Why should you practice personal cleanliness and good grooming?
2. You are gaining more weight than usual. What can you do about it?

SCIENCE
In Your World

Healthful Eating

Do you ever hear your stomach growling? It's letting you know it's time to eat. What do you eat? How often? And how much? The answers to these questions may be different for everyone, but everyone needs food for the same reasons. Food helps you grow and stay warm, and keeps your body working properly. Be sure to eat a variety of the right kinds of food.

Have You Ever...

Separated Fat From Cream?

Pour heavy cream into a small container until it's half full. Place two clean marbles into the container and fasten the lid tightly. Then carefully shake the container until the liquid thickens. Remove the lid and observe the contents. What do you see? Why?

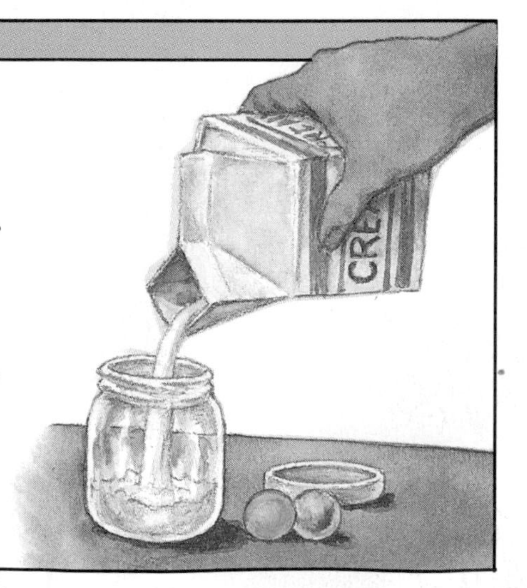

Nutrients

LESSON 1 GOALS
You will learn
- what nutrients your body needs.
- how these nutrients affect your body.
- what food sources provide these nutrients.

If you had to make a list of all the things you need to live, what would your list consist of? Probably near the top of your list would be food. Food would not be an easy thing to forget, because we all get hungry. But why exactly do we need food?

Food is needed by your body for energy. Your body also needs food for growth and repair of body tissue. Food meets these needs of your body because it contains **nutrients.** Although your body actually needs more than 50 nutrients to stay healthy, you will learn about 6 main kinds of nutrients your body needs in balanced amounts every day.

Proteins

Besides water, you have more proteins (PROH teenz) in your body than any other material. **Proteins** are nutrients your body needs for growth and repair of your body cells. Proteins also provide a source of energy for your body. Sources of protein are meat, fish, poultry, eggs, milk and dairy products, whole grains, and bean products.

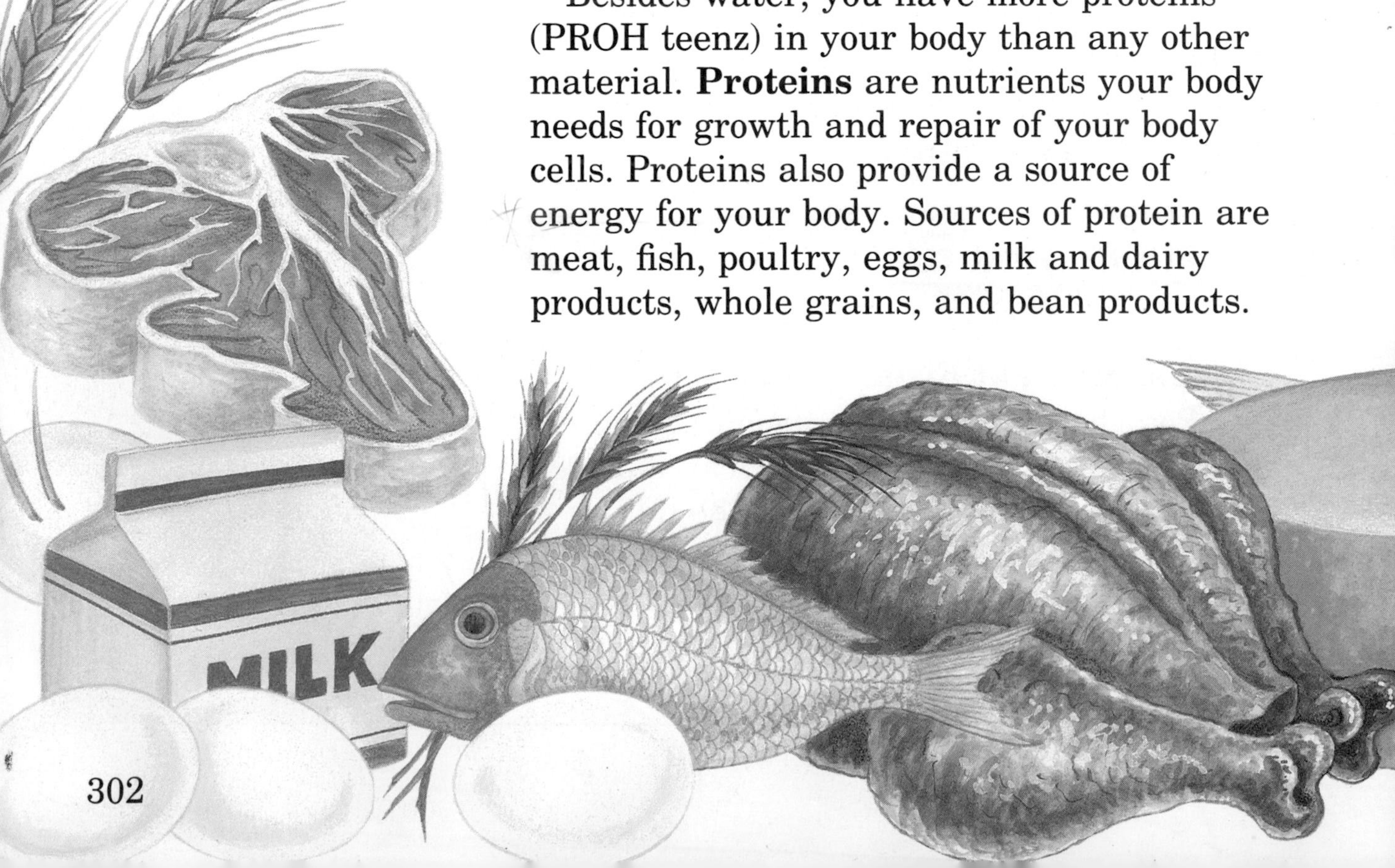

Carbohydrates

Your body's main source of energy is provided by **carbohydrates** (kar boh HI drayts). These nutrients supply energy for all your body functions.

The main carbohydrates in food are sugars and starches. Natural sugars, such as those in fruits, are easy to digest and quickly provide energy. Starches have to be broken down into sugars during digestion before they are used by the body. The best food sources of carbohydrates are whole grains, fruits, and vegetables.

Fats

Your body gets the most concentrated form of energy from **fats.** Your body uses fats to transport and store some vitamins. Fat deposits protect your organs and hold them in place. They also insulate your body. Foods that contain fats are butter, oils, margarine, nuts, seeds, meat, whole milk, eggs, and cheese.

ACTIVITY

You Can...

Test for Fat

Cut a 10 cm x 10 cm square from a brown paper bag. Rub some cooking oil on one edge. Rub some water on another edge. Watch what happens. Hold the paper up to the light. What do you see? Use this test to check foods for fat by rubbing a small amount of each food on the square. Try margarine, bread, milk, an apple, a potato chip, and bacon.

Vitamins

What should you do to be sure you get all the vitamins you need?

Your body could not grow properly without vitamins (VITE uh munz). **Vitamins** are nutrients that help your body use protein, fat, and carbohydrates. They help change fat and carbohydrates into energy. They help form bone and other body tissue. Vitamins also aid in building all your body parts and help regulate your digestion. Though some foods are high in certain vitamins, you must eat a variety of foods to be sure you get all the vitamins you need. Carrots and yams are good sources of vitamin A. Nuts and whole grains in cereal provide several forms of vitamin B.

Minerals

Minerals are nutrients that every part of your body needs. Minerals are important for strengthening your skeleton and making your teeth strong. They help keep your heart strong and your brain working as it should.

Your body needs about 16 different minerals for good nutrition. You need some minerals, such as calcium in milk and potassium in fruits and green leafy vegetables, in fairly large amounts. Other minerals, such as iron in meat, are needed only in small amounts.

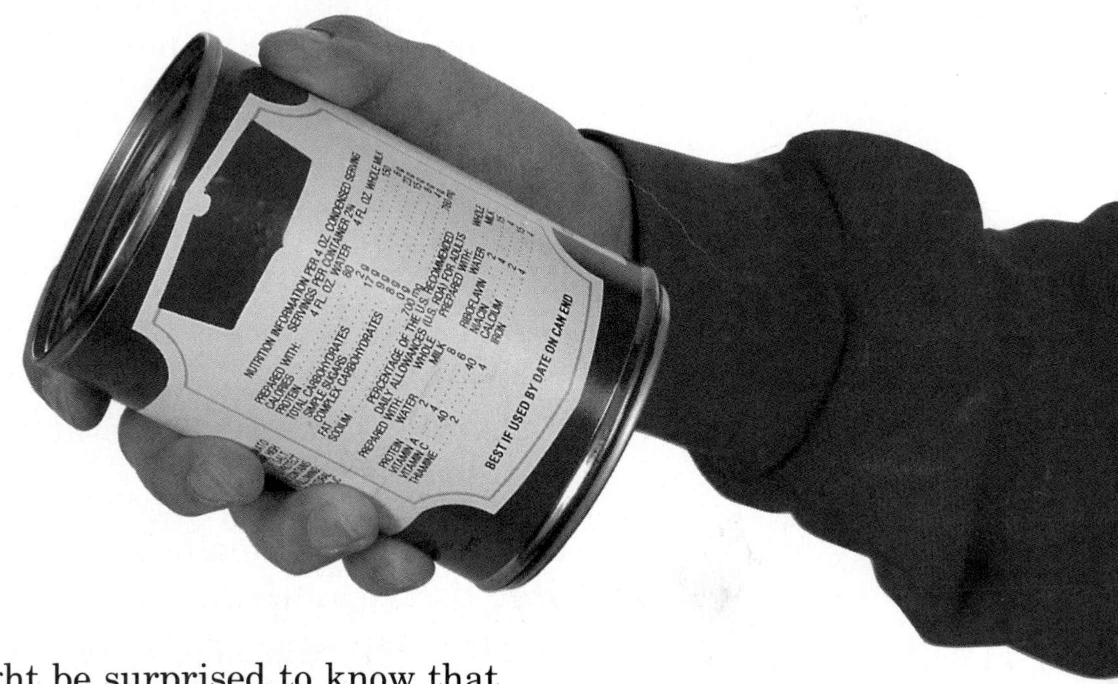

This food provides a variety of nutrients.

Water

You might be surprised to know that water makes up about two-thirds of your body weight. Every cell in your body needs water. As part of blood, water moves nutrients through your body. Water helps keep your body temperature normal and carries waste material out of your body.

You get water from many sources.

You can get water from many sources. You may drink water or get water in other beverages. You also get water in soups, fruits, and vegetables.

What are some sources of water other than beverages?

Your body can't get along without certain things. So, the next time your stomach growls, stop and consider what your body needs. Eating is a lot more than just satisfying hunger.

Lesson Summary

- Your body needs six main nutrients—proteins, carbohydrates, fats, vitamins, minerals, and water.
- The six main nutrients provide materials for energy and the growth and repair of your body cells.
- You get the nutrients your body needs from eating a variety of foods.

Lesson Review

1. What foods provide protein?
2. What nutrient is the main source of energy for your body?

★3. How is fat used by the body?

How can you test for one kind of sugar?

What you need

test strips for sugar (Clinistix)
4 small cups
small spoon
honey
table sugar
water
apple juice
milk
pencil and paper

What to do

1. Put 1/2 spoonful of honey in a cup. Quickly dip a test strip in and out of the honey.
2. After 10 seconds, compare the color of the strip with the color chart that comes with the strips. Record your observations.
3. Mix 1/2 spoonful of table sugar with 1/2 cup of water. Quickly dip a new test strip in and out of the water. Repeat step 2.
4. Test juice, milk, and water. Record your observations.

What did you learn?

1. What changes did you observe on the test strip when the one kind of sugar (glucose) was present?
2. Which foods contained that kind of sugar?

Using what you learned

1. How would you test other foods for this sugar?
2. Infer other kinds of foods that might contain this sugar.

Food Groups and Nutrition

LESSON 2 GOALS
You will learn
- that five main food groups supply the nutrients your body needs.
- that you need daily servings from each food group.

Have you ever heard someone say, "You are what you eat"? What do you think these words mean? Do the words say anything to you about how important it is for you to choose your foods wisely before you eat?

The food you eat each day is your **diet.** Imagine what your body would be like if you had only candy and potato chips to eat. Not only would your diet be very boring, but your body would miss out on some very important nutrients.

We can divide the foods we eat into five groups according to the main nutrients they contain. The groups are milk, meat, fruit-vegetable, grain, and combination. You can use the food groups to plan a balanced diet.

Balance the foods in your diet.

Food Groups

Foods in the **milk group** contain these important nutrients—calcium, vitamin B_2, and protein. You need these nutrients for strong bones and teeth, healthy skin, and good vision. Foods in this group include milk, yogurt, pudding, cheese, and ice cream. To have a balanced diet, you should eat or drink three servings from the milk group each day.

Enjoying a healthful snack

Foods in the **meat group** contain protein, niacin, iron, and vitamin B_1. You need these nutrients for muscle, bone, blood cells, and healthy skin and nerves. This food group includes cooked lean meat, fish and poultry, eggs, and peanut butter. Your body needs two servings a day from this group.

SCIENCE AND . . .
Math

Antonio helps his dad make pizza. Today he noticed that five out of ten customers ordered pepperoni on the pizza. What is the probability that the next customer will order pepperoni?

A. 1 out of 10
B. 1 out of 5
C. 1 out of 2

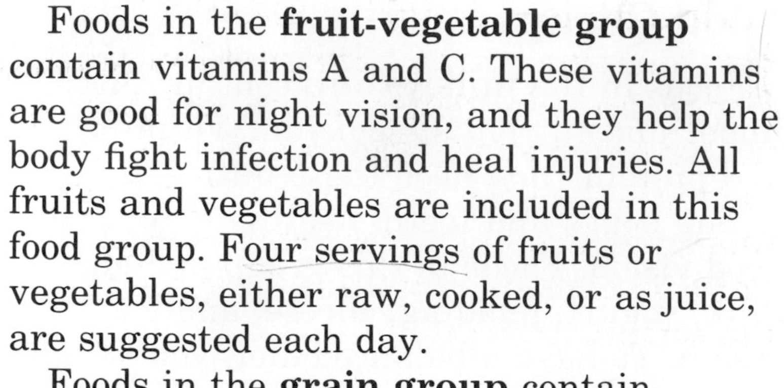

Foods in the **fruit-vegetable group** contain vitamins A and C. These vitamins are good for night vision, and they help the body fight infection and heal injuries. All fruits and vegetables are included in this food group. Four servings of fruits or vegetables, either raw, cooked, or as juice, are suggested each day.

Foods in the **grain group** contain carbohydrates, vitamin B_1, iron, and niacin. Your body needs these nutrients for energy and a healthy nervous system. This food group includes bread, cereal, pasta, and rice. You should include four servings from the grain group in your diet each day.

Oodles of noodles

The fifth food group is called the **combination group** because it contains foods from more than one food group. Foods in this group supply the same nutrients as the foods they contain. Some of the foods in this group are soup, stew, chili, and pizza. Can you explain why these foods are in the combination group?

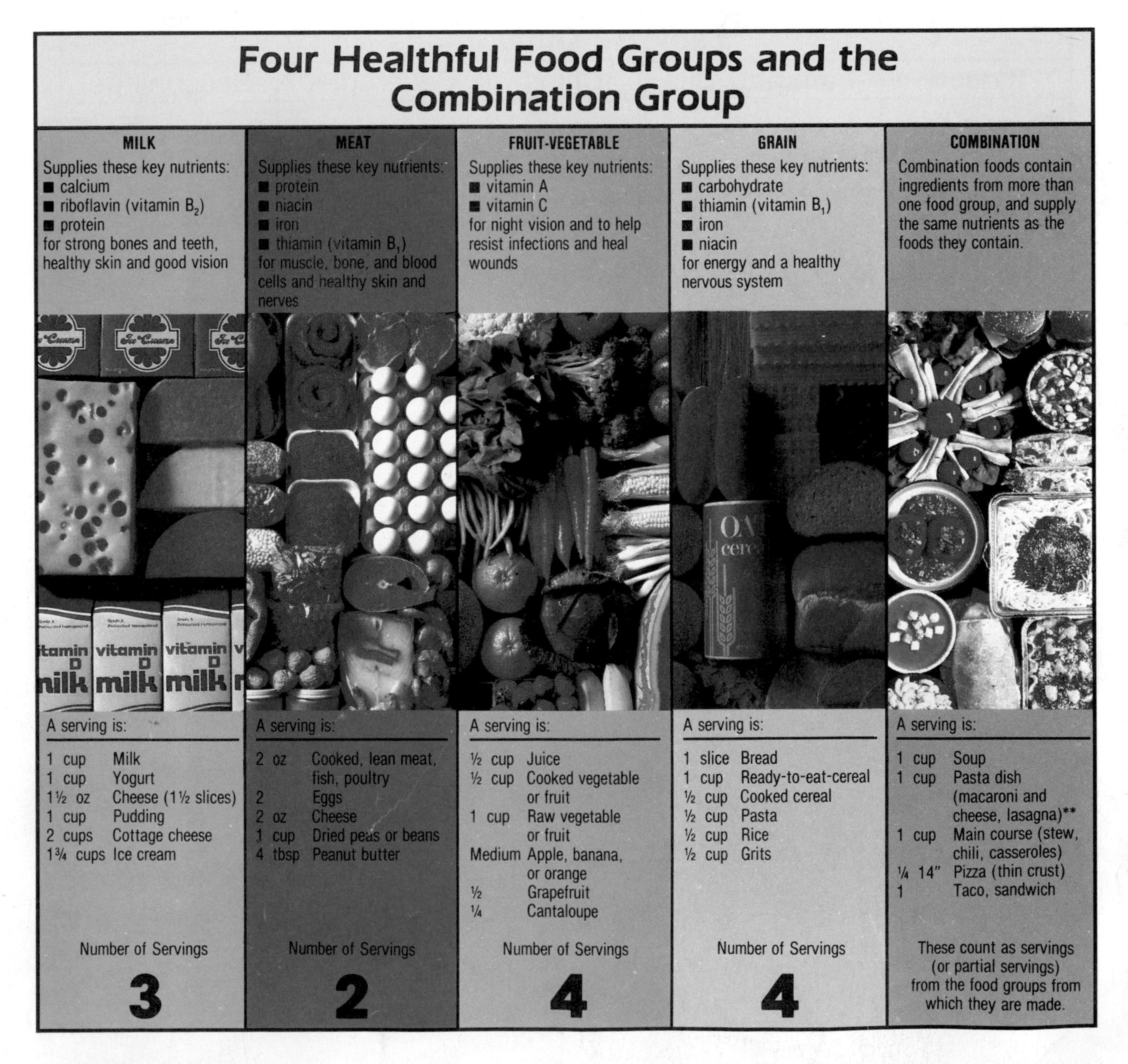

Four Healthful Food Groups and the Combination Group

MILK	MEAT	FRUIT-VEGETABLE	GRAIN	COMBINATION
Supplies these key nutrients: ■ calcium ■ riboflavin (vitamin B_2) ■ protein for strong bones and teeth, healthy skin and good vision	Supplies these key nutrients: ■ protein ■ niacin ■ iron ■ thiamin (vitamin B_1) for muscle, bone, and blood cells and healthy skin and nerves	Supplies these key nutrients: ■ vitamin A ■ vitamin C for night vision and to help resist infections and heal wounds	Supplies these key nutrients: ■ carbohydrate ■ thiamin (vitamin B_1) ■ iron ■ niacin for energy and a healthy nervous system	Combination foods contain ingredients from more than one food group, and supply the same nutrients as the foods they contain.
A serving is: 1 cup Milk 1 cup Yogurt 1½ oz Cheese (1½ slices) 1 cup Pudding 2 cups Cottage cheese 1¾ cups Ice cream	A serving is: 2 oz Cooked, lean meat, fish, poultry 2 Eggs 2 oz Cheese 1 cup Dried peas or beans 4 tbsp Peanut butter	A serving is: ½ cup Juice ½ cup Cooked vegetable or fruit 1 cup Raw vegetable or fruit Medium Apple, banana, or orange ½ Grapefruit ¼ Cantaloupe	A serving is: 1 slice Bread 1 cup Ready-to-eat-cereal ½ cup Cooked cereal ½ cup Pasta ½ cup Rice ½ cup Grits	A serving is: 1 cup Soup 1 cup Pasta dish (macaroni and cheese, lasagna)** 1 cup Main course (stew, chili, casseroles) ¼ 14" Pizza (thin crust) 1 Taco, sandwich
Number of Servings **3**	Number of Servings **2**	Number of Servings **4**	Number of Servings **4**	These count as servings (or partial servings) from the food groups from which they are made.

Some foods don't supply the body with needed nutrients, so they are placed in a separate group called the others group. These foods should not be used as replacements for foods in the five main food groups. You can look at the table on this page for examples of foods in the others group.

Table 1 Other Foods

Sweets	Fats and Oils	Chips	Beverages	Seasonings
brownies	butter	corn	coffee	catsup
cookies	margarine	potato	soft drinks	mustard
candy	cream	pretzels	tea	olives
jelly, jam	gravy	tortilla chips	fruit-flavored drinks	pickles
sugar	mayonnaise			sauces
doughnuts	salad dressing			spices

Some foods are high in Calories and low in nutrients.

Now that you know more about the food groups and why you need a variety of foods, you can make better choices about the foods you eat. Some foods will taste better to you than others. Some foods will always be your favorites. But the important thing to remember is that you just might like a food if you try it! Remember, "you are what you eat." What you eat makes a big difference.

Good food, good times, good health

Lesson Summary

- Servings from the milk group, the meat group, the fruit-vegetable group, and the grain group provide the nutrients your body needs. Some of these nutrients can be provided by eating foods from the combination group.
- Each day you need three servings from the milk group, two from the meat group, and four servings from both the fruit-vegetable and the grain group.

Lesson Review

1. Which food group is a good source of vitamins A and C?
2. How many servings should you eat each day from the grain group?

★3. Why should you be careful not to overeat foods such as candy, potato chips, and doughnuts?

Use Application Activity on pages 373, 374.

How can you plan a balanced diet?

What you need

diet plan chart
Food Group Chart
pencil and paper

What to do

1. Look at the chart provided by your teacher. The first column lists the 4 healthful food groups.
2. Study the menus that are given for breakfast, lunch, and dinner.
3. Use the Food Group Chart in your textbook. Make your own menus for these meals. Make sure you have a balanced diet.

What did you learn?

1. If a person does not want to eat meat or fish, what can they eat instead?
2. Do you need more servings from the fruit-vegetable group or the meat group to have a balanced diet?

Using what you learned

1. In what food group would each of the following belong?
   pizza dough
   raisins
   dill pickle
   walnuts
2. Why is the number of servings from each food group important in a balanced diet?
3. If a person did not have enough servings from the milk group, what could they have for a snack?

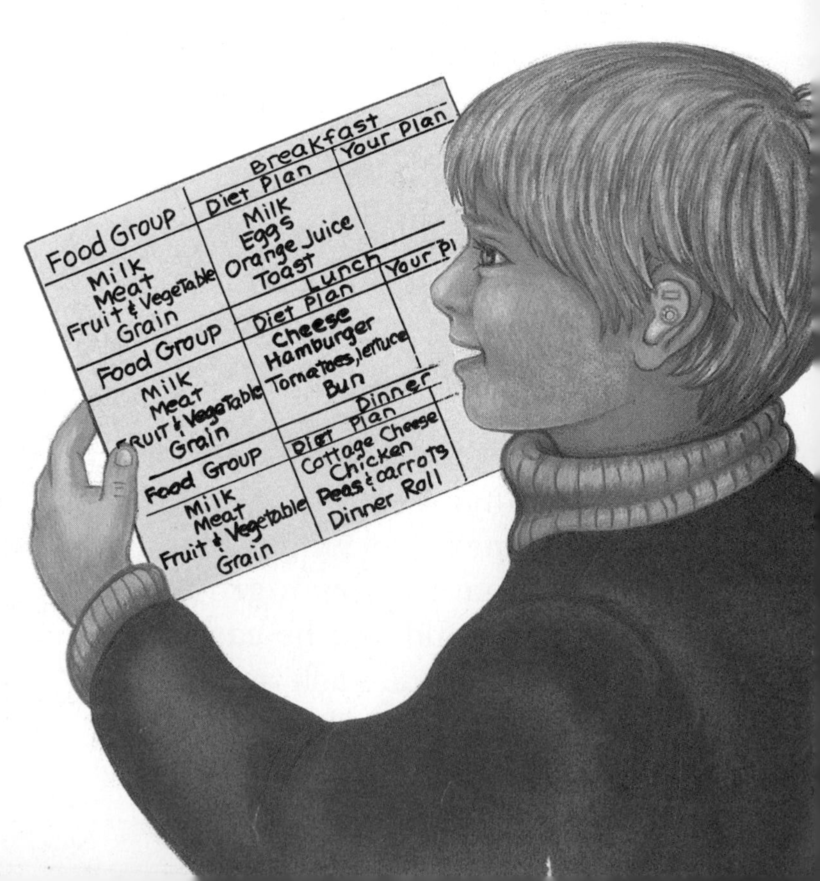

I WANT TO KNOW ABOUT...

Tooth Paint

You may already be familiar with finger paint, nail polish, hair dye, and face paint. But do you know that there is also a tooth paint? Tooth paint is a plastic sealer that prevents tooth decay.

Children between 5 and 17 years old get most of their cavities in the top surfaces of their back teeth. The back teeth, or molars, are used for chewing. These teeth have many small grooves, which trap food and germs. When sealer is painted on the chewing, or top, surfaces of back molars, it forms a protective covering.

To seal teeth, a dentist first cleans and dries the teeth. Next, he or she rubs the teeth with a solution of weak acid. After one minute, the acid is rinsed off, and the teeth are dried again. Then, the dentist paints the sealer on the teeth. In only one-half hour, all four back molars can be sealed.

The treatment is simple and painless. The sealer usually lasts up to five years, but the sealed teeth should be checked for cracks or worn spots at regular checkups.

Maybe someday soon you will add a new item to your list of good dental care habits. Besides eating a balanced diet, brushing, flossing, and getting regular checkups, you may also add tooth painting!

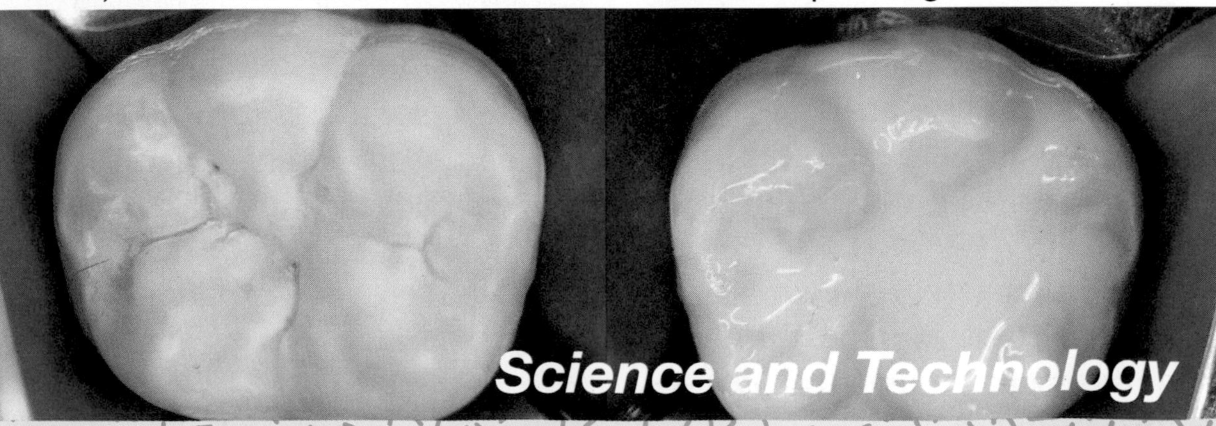

Science and Technology

CHAPTER REVIEW 14

Summary

Lesson 1

- Your body needs six main nutrients—protein, carbohydrates, fats, vitamins, minerals, and water.
- The six main nutrients provide materials for energy and the growth and repair of your body cells.
- You get the nutrients you need from eating a variety of foods.

Lesson 2

- Servings from the milk group, the meat group, the fruit-vegetable group, the grain group, and the combination group provide nutrients needed.
- Each day you need three servings from the milk group, two from the meat group, and four from both the fruit-vegetable and the grain group.

Science Words

Fill in the blank with the correct word or words from the list.

nutrients **protein**
carbohydrates **vitamins**
minerals **fats**
meat group **grain group**
fruit-vegetable group **milk group**
diet
combination group

1. Nutrients that supply energy for all your body functions are called ___.
2. Nutrients your body needs for growth and repair of your body cells are ___.
3. Nutrients that transport and store some vitamins are ___.
4. Your body uses proteins, fats, and carbohydrates with the help of ___.
5. The food you eat each day is your ___.

Questions

Recalling Ideas

Correctly complete each of the following sentences.

1. Meat, fish, and dairy products are good sources of
   (a) water. (c) fats.
   (b) proteins. (d) carbohydrates.
2. Your body gets its main source of energy from
   (a) carbohydrates. (c) fats.
   (b) proteins. (d) minerals.
3. Your body could not grow properly without
   (a) carbohydrates. (c) proteins.
   (b) vitamins. (d) fats.
4. Foods in the milk group contain
   (a) calcium. (c) protein.
   (b) vitamin B_2. (d) all of these

5. Fish, poultry, eggs, and peanut butter are foods in the ___ group.
   (a) combination (c) milk
   (b) grain (d) meat

Understanding Ideas

Answer the following questions using complete sentences.

1. What are the best food sources of carbohydrates?
2. How can you be sure you get all of the vitamins you need?
3. Why is water an important nutrient?
4. What are the six main nutrients?
5. How are the foods you eat grouped, and what are the groups?
6. What is the combination group and what sorts of foods does it contain?

Thinking Critically

Think about what you have learned in this chapter. Answer the following questions using complete sentences.

1. Why is it important to have fat deposits in your body?
2. Describe the daily components of a good diet.

SCIENCE
In Your World
Fleet

Drugs

If you've ever been sick, you've probably taken medicine that your parent bought in the drugstore or that your doctor prescribed. These are legal drugs that will help to make you well. However, taking illegal drugs won't make you feel good. They can only harm you.

Just Said No to Drugs?

Would you know what to do if someone offered you an illegal drug? Try this role-playing experiment. Imagine you are with a group of friends. One of them suggests that everyone try some "stuff to feel good." You realize you are being offered an illegal drug. How would you refuse the offer and remove yourself from the situation? What would you do if you found out that your best friend was using drugs?

Learning About Legal Drugs

LESSON 1 GOALS
You will learn
- what a drug is.
- the difference between prescription and over-the-counter drugs.
- what precautions should be taken before using any medicine.

What do aspirin, wine, cigarettes, coffee, and cough medicine have in common? Each of these contains a drug. A **drug** is a chemical that changes the way a person feels, thinks, or acts. *Drugs* is a very general term that includes many different types of chemicals.

You may not realize it, but you are bombarded with information about drugs. Try an experiment. Tonight, look through some magazines and newspapers, or spend a few minutes watching television. Count how many times you see an article or an advertisement or a program about drugs. More than likely, you probably aren't aware of all the items that actually are drugs. After reading this chapter, do the experiment again. Your count will probably increase. Your new knowledge about drugs will make you more aware and enable you to make healthful decisions.

It's important to know about different types of drugs.

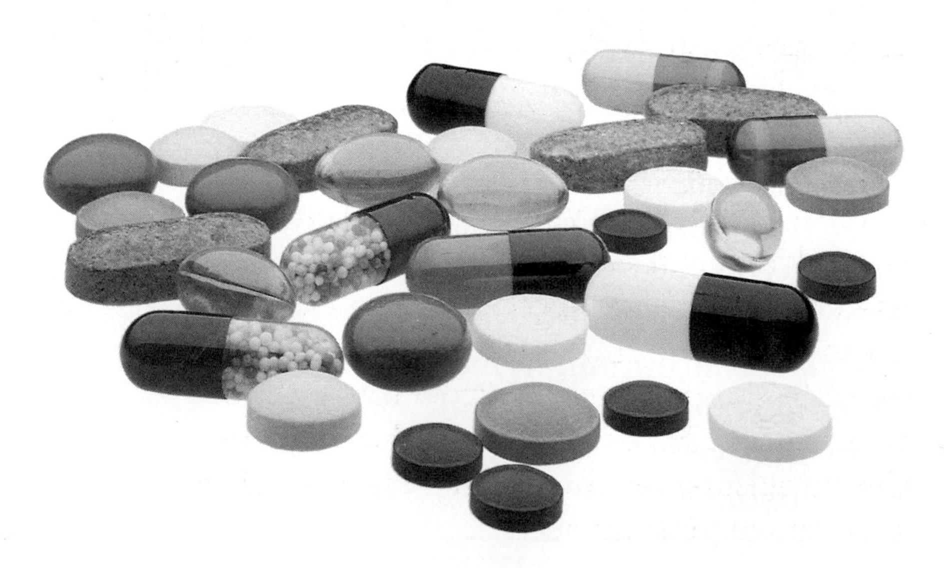

Legal Drugs

Medicines are drugs. The cough medicine that your doctor may give you for a bad cough is a legal drug. A **legal drug** is a drug that you are allowed to take by law. You can get legal drugs through your doctor, or you can buy them at a store. Examples of legal drugs are aspirin, cold medicines, and first-aid ointments.

Legal drugs are used to treat illnesses.

When used properly, these drugs can be helpful in keeping you healthy. But even legal drugs can be harmful if they aren't used according to directions. If a doctor gives you a cough medicine for your terrible cough, it will help you get rid of your cough. However, if you don't follow the directions on the bottle, the medicine may cause you harm.

When can legal drugs be harmful?

Prescription Drugs

A medicine that your doctor orders for you is a **prescription drug.** A prescription drug is a legal drug. It can be ordered only by a doctor, who will order it for you to pick up at a store. Prescription drugs can be helpful when taken according to the directions. They may be used to prevent or cure disease or to ease pain.

It's important that you take prescription drugs only with your parent's or guardian's permission. Since these drugs are prescribed for a particular person, it is important to never take another person's prescription drug.

Only a parent or guardian should give medicine to you.

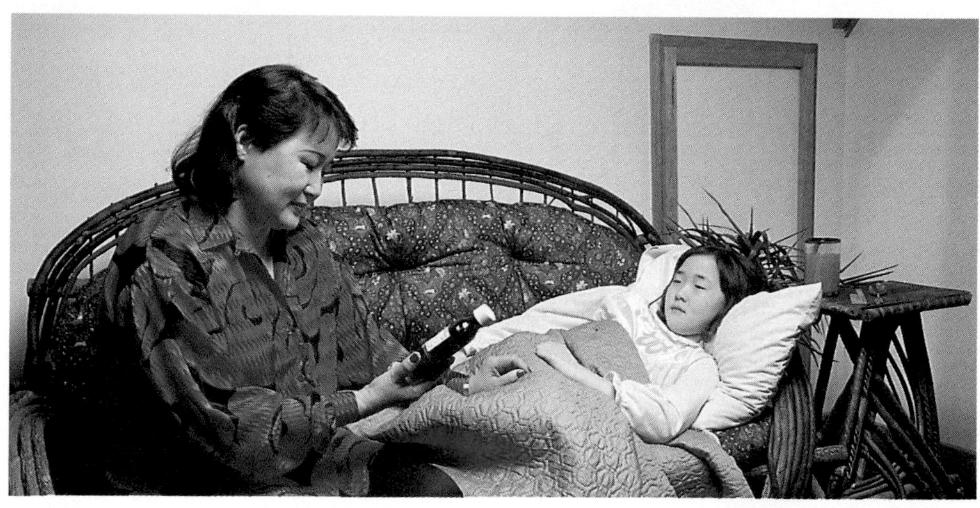

Over-the-Counter Drugs

Have you ever noticed a section in some stores where you can buy such items as aspirin or cold medicines? These types of medicines are called over-the-counter (OTC) drugs. **OTC drugs** can be purchased without a doctor's order. Some common OTC drugs are aspirin, skin ointment, and cold medicine.

Your parent can make wise choices about OTC drugs.

Like prescription drugs, OTC drugs are helpful as long as they are used according to their directions. You should not use OTC drugs without the permission of your parent or guardian.

Here are some other important things to remember when taking prescription or OTC drugs.

- Read the label of the drug carefully.
- Follow the directions exactly.
- Do not use a drug for longer that it is to be used.
- Make sure the seal around an OTC drug is not broken. A broken seal may mean the drug has been damaged or changed by someone.
- Ask your doctor before using more than one drug. Some drugs can cause harmful effects when used in combination with other drugs.
- Be aware of any side effects a drug may cause. For example, some sinus medications may cause you to become sleepy.

Young children are given medicine to prevent illness.

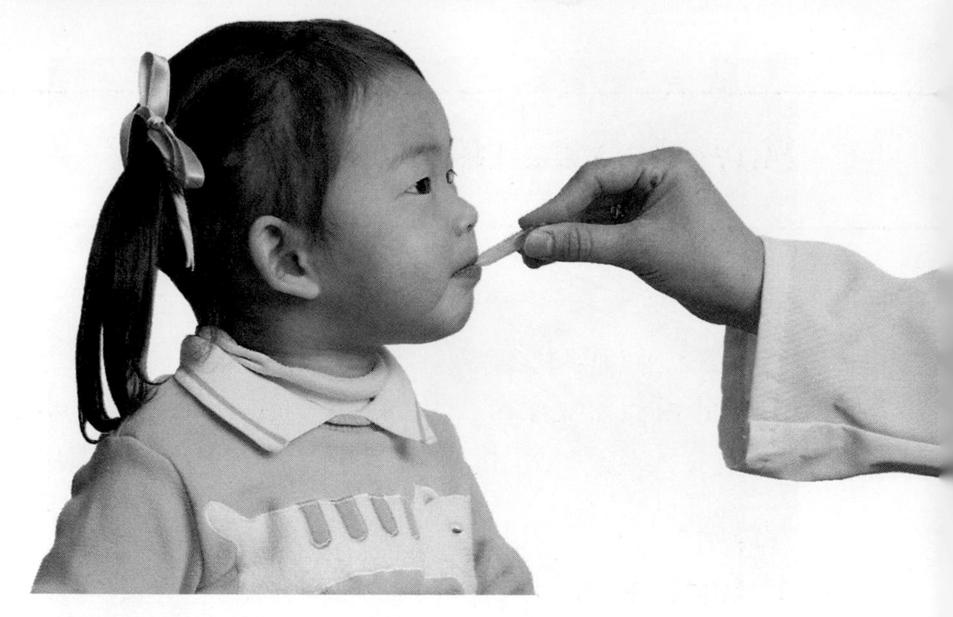

Legal drugs can improve the quality of our lives. When used properly, they can cure, treat, or prevent illnesses. The best way for you to avoid harm from using drugs is to never take drugs without the permission of your parent or guardian.

Lesson Summary

- A drug is a chemical that changes the way a person feels, thinks, or acts.
- Prescription drugs can be purchased when ordered by a doctor, while OTC drugs can be purchased as needed.
- It is extremely important to read the label of any medicine carefully and to follow the directions exactly.

Lesson Review

1. List some examples of legal drugs.
2. Name four things to remember when taking prescription and OTC drugs.
★**3.** Why can it be dangerous not to know the possible side effects of a drug you are taking?

How do you read drug labels?

What you need
pencil and paper

What to do
1. Look at the drug labels on this page. One label is for an over-the-counter drug. The other one is a prescription drug label. Read them carefully.
2. Answer the questions about the labels.

RELIEVES COUGHS AND UPPER CHEST CONGESTION DUE TO COLDS, TO HELP YOU REST.

Directions for Use (2 years and older):
2-6 years – 1 teaspoonful
6-12 years – 2 teaspoonfuls
ADULTS (12 years and older) – 4 teaspoonfuls
Repeat every 4 hours as needed.

WARNING: Do not use for persistent cough, such as with asthma, emphysema, smoking, or cough with excessive phlegm, unless directed by a doctor. Persistent cough may indicate a serious condition. If cough persists past one week, recurs, or is accompanied by fever, rash or persistent headache, consult a doctor. In case of accidental overdose, seek professional assistance or contact a poison control center immediately. If you are pregnant or nursing a baby, seek the advice of a health professional before using this product.
As with all medicines, keep out of children's reach.
ACTIVE INGREDIENTS per tsp. (5 ml.): Dextromethorphan Hydrobromide 3.5 mg., Guaifenesin (Glyceryl Guaiacolate) 50 mg.

EXP. DATE: SEE BOTTOM. STORE AT ROOM TEMPERATURE. AVOID EXCESSIVE HEAT.

Richardson-Vicks Inc.
Health Care Products Division, Wilton, CT 06897 U.S.A.

BRINKMAN
PHARMACY & HOME HEALTH CA
SOUTH STATE ST. 882-2375 WESTERVILLE, OH
10000 Dr. Smith
Joe Doe 7-28-88rm
Take 1 to 2 tablets every 4hours for pain
Acetaminophen/ Codeine #3

What did you learn?

Over-the-counter Drug Label

1. How much of this medicine should be given to a nine-year-old and how often?
2. What should you do if someone takes an accidental overdose of this medicine?
3. What is the main use for this drug?

Prescription Drug Label

1. What is the patient's name?
2. What is the name of the doctor who prescribed the drug?
3. How much of the drug is to be taken at a time?
4. Why was this drug prescribed?

Using what you learned
1. Why should you only take drugs given to you by a parent or other trusted adult?
2. Why have drug companies put child-proof caps on drug containers?

Other Kinds of Legal Drugs

LESSON 2 GOALS
You will learn
- that caffeine is a drug found in some foods and drinks.
- that tobacco contains harmful drugs.
- the effects of using alcohol.

Cola, tea, coffee, and chocolate all have one thing in common—caffeine. Many of the foods you eat and beverages you drink contain this drug. In this lesson, you will learn about three legal drugs—caffeine, nicotine, and alcohol. Although legal, these drugs may be harmful to your health. However, it is important to remember that nicotine and alcohol are illegal for people your age to buy or use.

Food and drinks with caffeine

Caffeine

The next time you're in your favorite grocery store, walk up and down the aisles to see how many items have labels that say "caffeine-free." Why should you know whether or not products contain caffeine (ka FEEN)? **Caffeine** is a drug that speeds up the nervous system and helps keep a person awake.

Coffee contains caffeine. Caffeine can also be found in some medicines, chocolate, some teas, and many soft drinks. It would be a good decision to limit drinks and candy that contain caffeine and use "caffeine-free" products. How many kinds of food or drink containing caffeine do you have in your diet?

Nicotine

Do you think of the word *drug* when you see someone light a cigarette, a pipe, or a cigar? Perhaps not. However, tobacco is used in these products, and tobacco contains many harmful drugs. One harmful drug in tobacco is nicotine (NIHK uh teen). **Nicotine** is a drug that speeds up the nervous system, increasing the heart rate and blood pressure. In other words, the heart must work harder than it should.

Tobacco contains harmful drugs.

How is smokeless tobacco harmful?

The smoke from cigarettes contains many other harmful substances. One of these is tar. Tar is a sticky substance that irritates and clogs the lungs. Tar increases the chances of lung cancer and other respiratory problems.

When you think of tobacco, you probably think of cigarettes or cigars. However, there are other forms of tobacco. Smokeless tobacco is either chewed or placed between the cheek and gums. Do you think that tobacco is harmful only when it is smoked? Many people do, but this isn't true. Smokeless tobacco can cause many harmful effects to the body—especially to the mouth and teeth. Sugar in the tobacco can cause cavities. Over a period of time, gums pull away from the teeth. This can cause the teeth to loosen and fall out. Smokeless tobacco can also cause cancer. People who use smokeless tobacco risk getting throat and mouth cancer.

Smokeless tobacco

Alcohol

Another substance that many people do not think of as a drug is alcohol. However, alcohol is a drug, and it can harm your health. Drinking alcohol is illegal for someone your age. The age for legally drinking alcohol is 21 in every state.

Alcohol is a major cause of accidents.

How does alcohol affect the body?

Alcohol is a drug that is found in beer, wine, whiskey, and other kinds of beverages. Alcohol slows down the brain and other parts of the nervous system. It dulls the senses of sight, taste, touch, smell, and hearing. Shortly after a person begins to drink, the alcohol reaches the bloodstream and travels to all parts of the body. Alcohol can cause blurred vision, errors in thinking and judgement, slowed reaction time, and loss of muscle control. Your senses are very important in helping to keep you safe. When senses are dulled, the chances of having an accident increase. Long-term use of alcohol can cause damage to the liver, stomach, throat, and other body organs.

Although the drugs discussed in this lesson are legal drugs, the use of them can cause much harm to the body and to many areas of a person's life. Being informed about these drugs and their potential effects allows you to make intelligent decisions about their use.

Lesson Summary

- Caffeine is a drug found in coffee, tea, chocolate, and some soft drinks.
- Nicotine and tar are harmful substances in tobacco.
- Alcohol use can cause damage to body organs and can dull all the senses.

Lesson Review

1. How does nicotine affect the heart?
2. Name three possible harmful effects of using smokeless tobacco.
3. What might be a side effect of taking medicine that contains caffeine?
★4. Give at least two examples of how alcohol use could cause an accident.

How is advertising used to sell drugs?

What you need
magazines
paper and pencil

What to do
1. Magazine, newspaper, radio, and TV ads are used to get people to buy things. It's important to look at ads carefully and to see how they try to influence people.
2. Use 3 or 4 magazines. Look for ads for OTC drugs, alcohol, and tobacco. Make a table like the one shown. Record the information from each ad in the table.

Advertisement Information
Name of magazine
Type of drug advertised
Description of picture
Slogans used
Warnings provided

What did you learn?
1. What kinds of pictures do the ads for smoking show?
2. What kinds of warnings are there in the ads for smoking?
3. What kinds of pictures do the ads for alcohol show?

Using what you learned
1. How are the pictures in the ads supposed to make people feel? What messages are they trying to send?
2. What is the purpose of advertising? What do the companies who placed the ads want to do?
3. Where have you seen ads against smoking and drinking? Who pays for those ads?

Learning About Illegal Drugs

LESSON 3 GOALS
You will learn
- why certain drugs are illegal.
- what marijuana, cocaine, and crack are and how they affect the body when used.

Some drugs are illegal, meaning that by law you are not allowed to buy, sell, or use them. **Illegal drugs** are ones that can cause harm to you and to others, even after taking them only one time. Some examples of illegal drugs are marijuana and cocaine.

Some drugs are illegal because they are dangerous to those using them as well as to others. They can damage a person's body and mind. Laws have been passed to protect you and others from these kinds of drugs.

Marijuana

Marijuana

Marijuana is an illegal and harmful drug made from the leaves, flowers, seeds, and stems of the hemp plant. People who use marijuana usually smoke dried parts of this plant. Marijuana contains hundreds of harmful substances. One of the most dangerous substances is **THC,** which causes many of marijuana's harmful effects. THC stays in the body for a long period of time.

A person who uses marijuana will cause harm to his or her body. The brain, the lungs, the heart, and other body organs are affected. Marijuana changes how a person thinks and learns. It causes the heart to beat faster and increases blood pressure. Marijuana can cause serious long-term effects. The body's ability to fight diseases is weakened. The chances of developing lung diseases, including cancer, are increased when a person uses marijuana.

Cocaine and Crack

Cocaine is another dangerous and illegal drug. It is a drug made from the leaves of the coca bush grown in South America. Cocaine can be mixed with other substances to form a drug called **crack.** Crack is also dangerous and illegal.

Cocaine and crack can have extremely harmful effects. They increase a person's heart rate and can interfere with the nerve signals to the heart. This can cause the heart to stop. A person can die even after using cocaine just once.

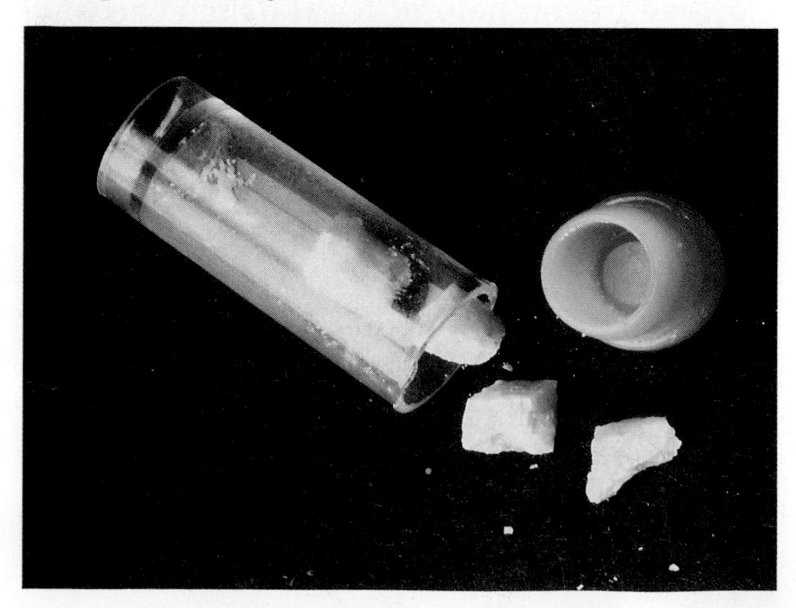

Crack is dangerous and illegal.

ACTIVITY

You Can...

Say NO to Drugs

You have heard this message many times since you have been old enough to watch TV and listen to the radio. Make a poster that shows how you can have a good time without using drugs. Display your poster in the hall so other students will see the message. How is advertising useful?

SCIENCE AND . . .
Math

Camille is decorating her "Just Say No" poster with a colorful border. The poster measures 60 cm wide and 90 cm long. How many cm of colored border does she need in all?

A. 180 cm
B. 120 cm
C. 240 cm
D. 300 cm

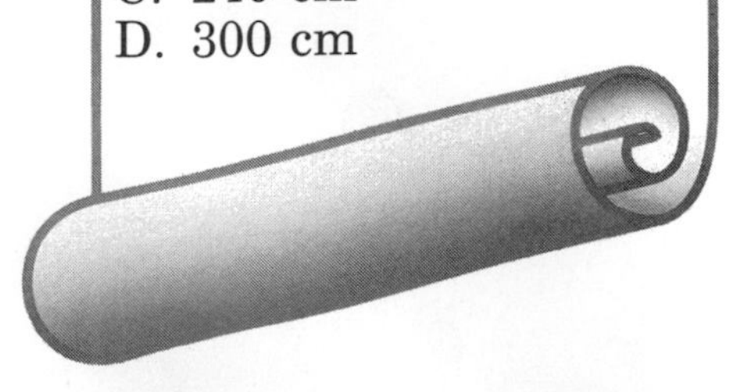

Looking back over what you've learned about drugs in this chapter, you have probably broadened your own definition of drugs. You now know that the term *drugs* can refer to legal drugs that are helpful, such as prescription and OTC medicines. However, even helpful legal drugs can have harmful effects if they are not used correctly. You also now know that the term *drugs* can refer to illegal drugs that are harmful and dangerous to use. People who use illegal drugs become dependent on them. Even though they would like to stop, they find it very difficult.

When a drug is illegal, it is against the law to make, buy, sell, use, or possess the drug. Punishments for breaking these laws can be very harsh.

Knowing everything you can about legal and illegal drugs can help you make wise decisions about your life. Most people know that illegal drugs are dangerous and do not use them. They do not want to harm their bodies.

Laws have been made to protect us from illegal drugs.

Lesson Summary

- Many drugs are illegal to protect people from the harm drugs can cause.
- Marijuana, cocaine, and crack are illegal drugs that can cause harmful effects, even death.

Lesson Review

1. What are the possible harmful long-term physical effects of using marijuana?
2. Why are cocaine and crack damaging to the heart?

★3. Why should you avoid using illegal drugs?

CHAPTER REVIEW 15

Summary

Lesson 1
- A drug is a chemical that changes the way a person feels, thinks, or acts.
- Prescription drugs must be ordered by a doctor, while OTC drugs can be purchased as needed.
- It's important to read the label on any medicine and follow directions.

Lesson 2
- Caffeine is a drug in coffee.
- Nicotine and tar are harmful substances in tobacco.
- Alcohol can cause damage to body organs and can dull all the senses.

Lesson 3
- Many drugs are illegal to protect people from the harm drugs can cause.
- Marijuana and crack are illegal drugs with harmful effects.

Science Words

Fill in the blank with the correct word or words from the list.

THC	**OTC drugs**	**nicotine**	**marijuana**
drug	**caffeine**	**alcohol**	**cocaine**
legal drug	**smokeless tobacco**	**illegal drugs**	**crack**
prescription drug			

1. A harmful drug found in tobacco is ____.
2. Drugs that you can purchase as needed are ____.
3. An illegal drug made from the hemp plant is ____.
4. A drug that you are allowed to take by law is a(n) ____.
5. An illegal drug made from the coca bush is ____.

6. Drugs that are dangerous and can damage a person's mind are ___.
7. The drug in beer is ___.
8. A chemical that changes the way a person acts is a(n) ___.
9. A drug that your doctor orders for you is a(n) ___.
10. A drug that is found in some coffee and tea is ___.

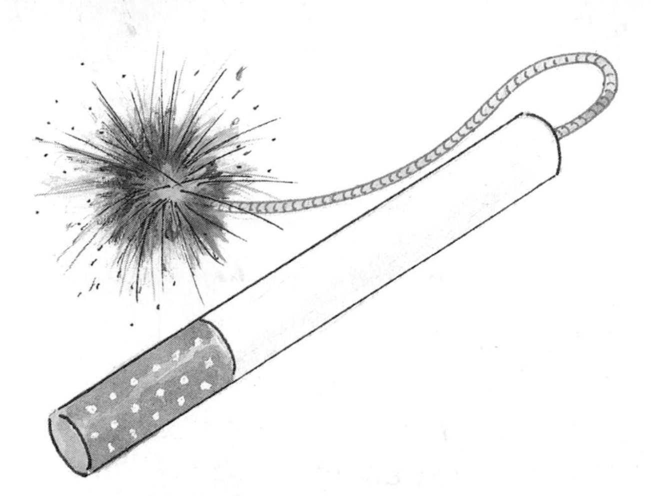

Questions

Recalling Ideas

Correctly complete each of the following sentences.

1. An illegal drug made from cocaine and other substances is
   (a) marijuana. (c) caffeine.
   (b) crack. (d) nicotine.
2. Blurred vision and slowed reaction time are caused by
   (a) crack. (c) alcohol.
   (b) marijuana. (d) aspirin.
3. Caffeine can be found in
   (a) coffee. (c) alcohol.
   (b) tobacco. (d) crack.
4. THC is a substance in
   (a) caffeine. (c) marijuana.
   (b) nicotine. (d) aspirin.
5. A drug that can cause your heart to stop is
   (a) aspirin. (c) alcohol.
   (b) nicotine. (d) crack.

Understanding Ideas

Answer the following questions using complete sentences.

1. How does caffeine affect your body?
2. List six precautions that should be taken before using any medicine.
3. How do marijuana, cocaine, and crack affect the body?

Thinking Critically

Think about what you have learned in this chapter. Answer the following questions using complete sentences.

1. What are the harmful substances in tobacco and what do they do to a person's body?
2. How is a cigarette similar to a cup of coffee?

UNIT REVIEW 4

Checking for Understanding

Write a short answer for each question or statement.

1. How does exercise improve physical health?
2. How does exercise improve mental health?
3. What are Calories?
4. How do unused Calories affect your body?
5. How can you maintain a healthful weight?
6. Why does your body need sleep every night?
7. Why is cleanliness important for good health?
8. How can you maintain a well-groomed appearance?
9. What nutrients does your body need?
10. How do nutrients affect your body?
11. What are the main food groups?
12. How many servings from each food group should you have every day?
13. What is the difference between an OTC drug and a prescription drug?
14. How can the use of nicotine and alcohol be harmful?
15. Why are some drugs illegal?
16. How can marijuana, cocaine, and crack be harmful?
17. Why is bathing important for healthy skin?
18. Why is brushing and flossing important for your teeth?
19. What do carbohydrates do for your body?
20. What vitamins are found in each of the food groups?
21. What products contain caffeine and how does it affect your body?

Recalling Activities

Write a short paragraph for each question or statement.

1. How many Calories do you use in one week?
2. Why wash your hands?
3. How can you test for sugar?
4. How can you plan a balanced diet?
5. How do you read drug labels?
6. How is advertising used to sell drugs?

Project Ideas

1. Design a survey and use it to find out how many adults participate in a lifetime sport. Use this survey to find out which sport is most popular among the adults you surveyed.
2. Using cotton swabs and agar-filled petri dishes, determine the presence of bacteria on teeth. Test several people at two hour intervals, then right after eating, immediately after brushing, and after using a mouthwash.

Books to Read

The Force Inside You by John Burstein, Coward-McMann, Inc.: NY, 1983.
Read about exercises for the body and mind.

Nutrition by Paul Thompson, Franklin Watts: NY, 1981.
Read this book and find out how the body uses nutrients.

What to Eat and Why by Ronald B. Fodor, William Morrow and Co.: NY, 1979.
Read this book and find out about the relationship between diet and health.

Application

Activities

There are many ways to learn about science. You may read or do activities. You use certain thinking skills to observe and record what happens. You use others to explain why something happens. Thinking skills are also used to solve problems. There activities were written to help you practice thinking skills. They were also written so you could use your imagination, to be creative, and to have fun.

TABLE OF CONTENTS

PROCESS SKILL MODELS

PROBLEM SOLVING ACTIVITIES

PROCESS SKILL MODEL

Measuring

Definition Measuring is finding out the size, volume, mass, weight, or temperature of an object. It is also finding out how long it takes for an event to happen. The object or event is compared to a unit of measure.

Example Mrs. Parker's students were doing an experiment with cloth, paper towels, and cardboard. Using a metric ruler, they measured and then cut a 25-cm strip of each material.

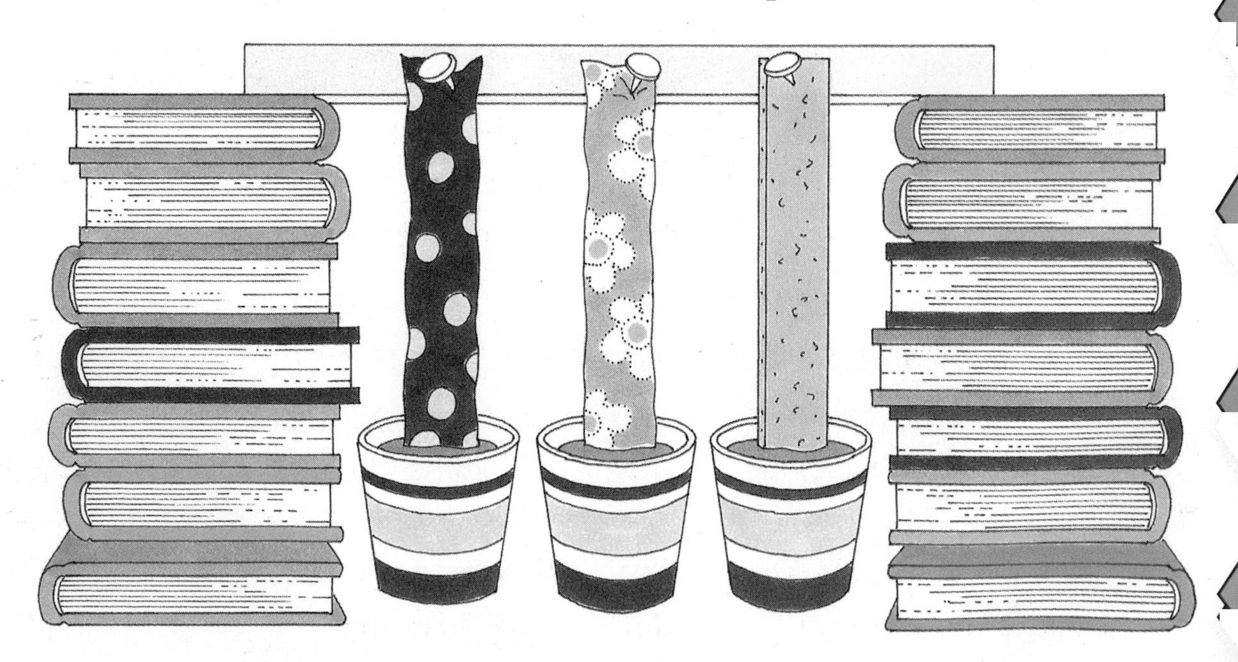

Next, they measured and poured 50 mL of colored water into each of the cups. They then lowered the strips into the colored water as shown. After three minutes, they removed the strips from the water. The distance the water moved up each material was recorded in the table below.

Movement of Water

Material	Distance Moved
cloth	13 cm
paper towel	5 cm
cardboard	1 cm

Practice

1. List what the students measured in this experiment. Include the units of measure that were used.
2. What else could you measure in this experiment?

PROCESS SKILL MODEL

Using Numbers

Definition Using numbers includes ordering, counting, adding, subtracting, multiplying, and dividing numbers.

Example Mrs. Reid's class built boats of aluminum foil. Each group of students placed paper clips in its boat, one at a time. They counted the number of paper clips their boat could hold before it sank. Each group redesigned its boat and tried again. Here are the results of each trial.

Cargo Boats

Group	Trial 1	Trial 2
Group 1	55 paper clips	78 paper clips
Group 2	40 paper clips	42 paper clips
Group 3	23 paper clips	63 paper clips
Group 4	101 paper clips	84 paper clips
Group 5	73 paper clips	91 paper clips

Practice

1. For the first trial, place the groups in order from the most paper clips to the least paper clips.
2. How many paper clips did the boat for group 1 hold for both trials?
3. Which group's boat held the most paper clips for both trials? How do you know?
4. Which group improved the most between trial 1 and trial 2? How did you find out?
5. What was the average number of paper clips the boat for group 5 held? Explain how you got your answer.

PROCESS SKILL MODEL

Predicting

Predicting is proposing possible outcomes of an event or experiment. Predictions are based on earlier observations and inferences.

Mrs. Morrell's students placed a 1-L bottle over a lighted candle. They found that it took 32 seconds for the candle to go out.

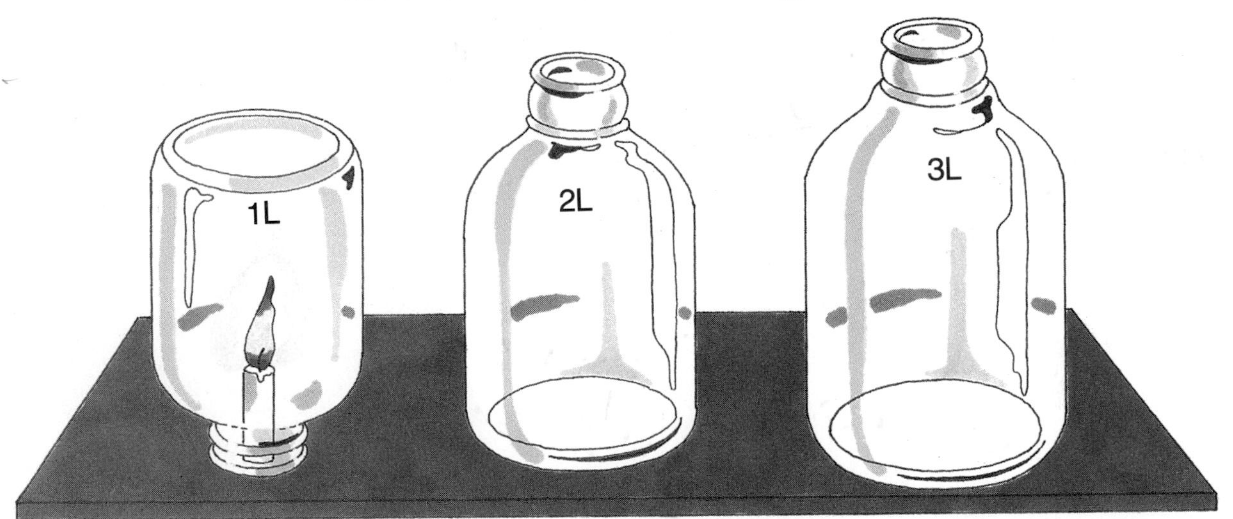

They predicted that a candle under a 2-L bottle would burn for 64 seconds. They tested their prediction and found it to be correct.

Burning Table

Bottle Size	Time
1 L	32 seconds
2 L	64 seconds

Practice

1. How long did the candle in the 1-L bottle burn? The 2-L bottle?
2. How do you think the students were able to make their predictions?
3. Predict how long you think a candle in a 3-L bottle would burn.
4. On what basis did you make your prediction?
5. Your teacher will set up an experiment to test your prediction. Make a table like the one below and record your results.

Burning Table

Bottle Size	Time
1 L	
2 L	
3 L	

PROCESS SKILL MODEL

Interpreting Data

Interpreting data is explaining the meaning of information that has been collected.

Ms. Foley's students were raising money for a science trip. They set up a fruit stand in the lunchroom. For the first week, they ordered 24 of each kind of fruit listed below.

Sara wanted to know which fruits were selling the most so the students could order enough for the next week. Tim recorded the number of each kind of fruit sold for the first week. The results are shown in the table below.

Fruits Sold

Fruit	First Week	Second Week
apples	24	
pears	18	
plums	6	
oranges	20	
peaches	10	

From the information collected, the class interpreted that the apples and oranges were the best sellers. In fact, they decided to order more apples for the second week!

Practice

1. Which fruit or fruits would you order less of? Why?
2. The following week, the students sold 32 apples, 17 pears, 2 plums, 27 oranges, and 12 peaches. Make a table like the one shown on the last page. Fill in the sales for the second week.
3. Interpret the sales data for both weeks. Do this for each kind of fruit. Which fruits would you buy more of for the third week?

PROCESS SKILL MODEL

Controlling Variables

Controlling variables is making sure that everything in an experiment stays the same except for one factor.

Karen left a piece of bread inside her desk on Friday. On Monday morning, mold was growing on it. She wondered if molds only grow in dark places. Karen decided to find out.

Friday came again. Karen got two pieces of fresh, moist white bread. She wrapped each in clear plastic wrap. One was left on top of her desk. The other was put inside her desk. On Monday, she found mold on both pieces of bread!

Practice 1. Which of the following factors, or variables, did Karen keep the same for both pieces of bread? Which did she change? Make a table like the one shown below and record your answers.

Bread Experiment

Variable	Same?	Changed?
kind of bread		
contact with air		
amount of moisture		
amount of light		
amount of time		
temperature		

2. Bill wants to see whether mold grows better on different kinds of bread. Tell how he should control the variables.
3. Does temperature affect mold? Tell how you would control the variables.
4. Why is it important to control the variables in an experiment?

PROCESS SKILL MODEL

Hypothesizing

Definition

Hypothesizing is making an educated guess about how or why something happens. A hypothesis can be tested to see if it is correct.

Example

Connie taught her dog to sit. Connie taught her dog by saying the word "sit." She rewarded him with food whenever he sat. The word "sit" was the stimulus. Sitting down was the desired response. Food was the reward. Based on her observations, Connie thought that a mouse could learn to find its way through a maze by using food as a reward. Connie decided to test her hypothesis.

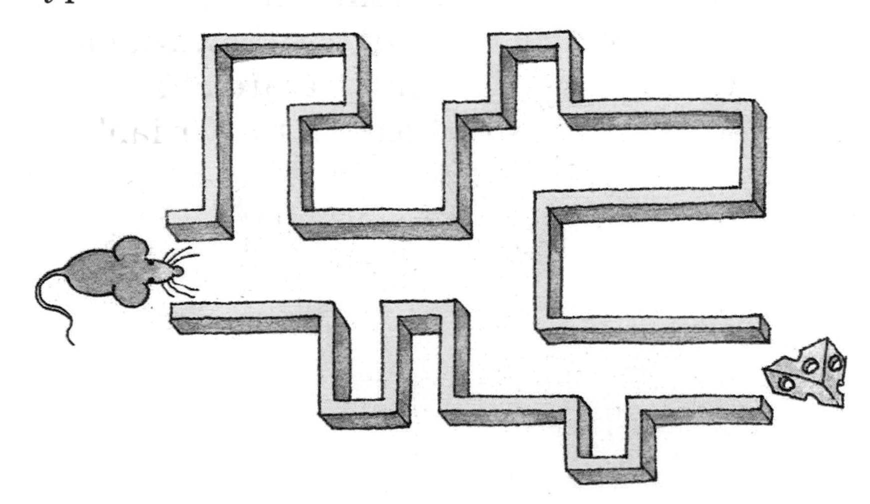

She obtained a mouse. She put it in a maze with food at the end. Connie recorded the time it took the mouse to find the food. She did this five times. Her results are shown in the table on the next page.

Mouse Maze

Trial	Time
1	5 minutes, 20 seconds
2	4 minutes, 15 seconds
3	3 minutes, 30 seconds
4	2 minutes, 58 seconds
5	1 minute, 35 seconds

Practice

1. What was Connie's hypothesis?
2. Was her hypothesis correct? Why or why not?
3. Write a hypothesis about what you think a certain animal can learn.
4. Tell how you would test your hypothesis.

PROBLEM SOLVING ACTIVITY

Mysterious Animals

Use after page 63.

Background On a vacation in a far-off land, you have discovered a strange animal. It is not exactly like anything you have ever seen. Yet, it has characteristics similar to animals that you have seen. As it turns out, each of your classmates has had a similar experience while on vacation! Each has discovered a mysterious animal.

Problem There are two problems included in this activity. The first problem is for you to recreate the mysterious animal that you have seen. You can use the materials listed below or any others you wish. The design for this animal will come from your imagination. However, your animal must have at least two characteristics used to classify animals as described in Chapter 3 of your textbook. The second problem is for you to classify the animals that your classmates invent.

Materials modeling clay • toothpicks • egg cartons • cardboard • paper • plastic containers • fur • feathers • other objects and materials you can find easily • glue

Solution

1. Create a data sheet for your animal. Draw a picture of it. List the characteristics you will include that will allow your animal to be classified. Be sure to include its classification. Build your animal.
2. Make the table as shown below. Observe the mystery animals of other students. List the characteristics of the animals. Then classify the animals.

Classification of Mystery Animals

Animal	Characteristics	Classification
1		
2		
3		
4		
5		

3. Compare your completed table with the information on the data sheets of the other students. How closely were you able to match the characteristics and classifications of their animals? Explain.

PROBLEM SOLVING ACTIVITY

Farmer Fred's Problem

Use after page 74.

Background

Each day many people eat cereal for breakfast. Companies that make cereals package them in boxes or bags. How do these companies get the ingredients that make up cereal? Cereals belong to the food group called grains. Farmers grow many of the ingredients found in cereals.

Problem

Farmer Fred wants to grow grains on his farm. He decides to plant seeds of the three kinds of grain that are most popular in breakfast cereals. If he grows these three grains, he will be able to make the most profit when selling the grains to cereal companies. He wonders which grains to grow. Help Farmer Fred by taking a survey of all students in your grade. Find out which breakfast cereal is preferred by each of the students. Make a table to list this information. Then determine which three cereals are most popular. Use the information to find out which grains Farmer Fred should grow.

Materials cereal boxes

Solution

1. Make a table like the one below to show the information you have found.

Cereal Survey

Student	Favorite Cereal
John	"Jolly Oats"
Erica	"Happy Day"
George	"Favorite Flakes"
Mary	"Jolly Oats"
Jane	"Kernel Cracky"
Aaron	"Happy Day"
Mike	"Jolly Oats"
Alice	"Favorite Flakes"
Constance	"Jolly Oats"
José	"Happy Day"

2. List the three cereals that are preferred by most students.
3. Using the answer to question 2, read the ingredients label on each box of cereal to find out which grains Farmer Fred should grow.

PROBLEM SOLVING ACTIVITY

A Classroom Garden

Use after page 100.

Background Planting a garden takes a lot of careful planning. You must choose the kinds of plants you want, and find out how and when to plant them. You must consider what plants grow well in the climate and soil conditions of your area. You must also decide what kind of garden you want to plant.

Problem Members of your class want to plant a garden. You and your class are to plan and design a classroom garden. The garden can be indoors or outdoors depending on your school setting. You must have at least four different kinds of plants in your garden.

Materials gardening books • packets of seeds • potting soil • watering can • gardening tools • drawing paper • container for garden

Solution

1. Plan your garden. Make a list of the things you should consider when planning a garden. What plants should you try?
2. Why did you choose these plants?
3. How can you decide which plants would be the best ones to grow?
4. How long will it take to produce each type of plant?
5. Find out what plants that are good to eat grow well in your area. Make a list.
6. Find out what time of year each type of plant should be planted and how long it takes until the plant is ready to eat.
7. Make up a planting and harvesting schedule.
8. Use gardening books to help you design the garden.
9. Make a scale drawing of the garden.
10. Plant the garden. Describe the results.

PROBLEM SOLVING ACTIVITY

Too Close for Comfort

Use after page 115.

Background You are the "wise person" on a planet called Hah. It is very, very hot and humid on Hah for part of the year. This happens when Hah comes close to its sun. Hah does this because its orbit is extremely elliptical. The people on Hah are sick and tired of their hot and humid summer. They have asked Hahnian scientists to change the orbit of their planet. But, Hahnian scientists do not know how to do this. They have come to you for help. You have something the Hahnian scientists do not have—a drawing that shows how to make an ellipse.

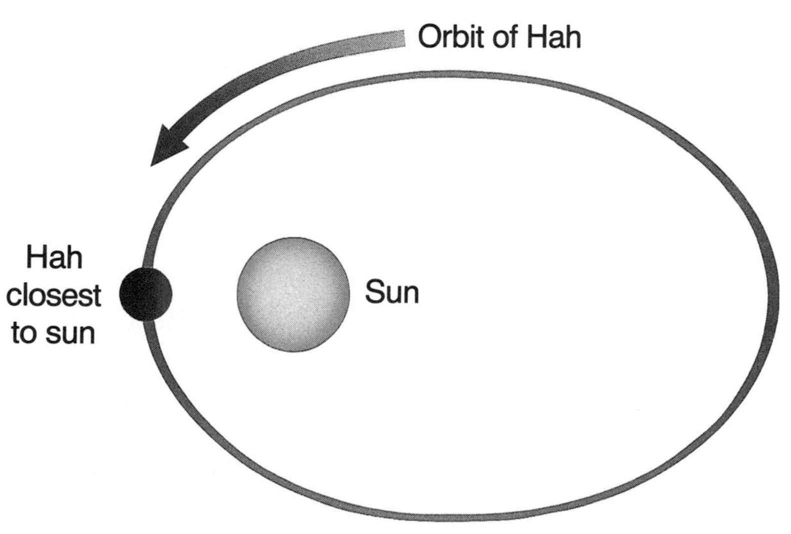

Problem Using the setup shown on the next page, change the orbit of Hah. The orbit should be changed so that Hah does not come as close to the sun. You cannot change the length of the string, and you cannot move the sun.

Materials large sheet of oak tag at least 44 cm x 44 cm • two push pins (one red, one white) • string • metric ruler

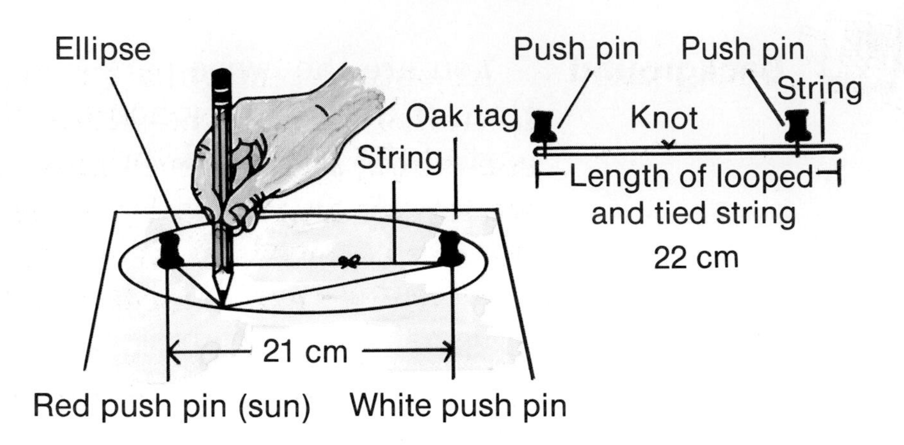

Solution

1. First, make an orbit for Hah that is similar to the one shown in the first drawing.
2. Now draw an orbit for Hah so that it is farther away from the sun at its closest point. The orbit must still be an ellipse. What factor did you change to produce a different orbit?
3. How did changing this factor affect the shape of the orbit of Hah?
4. How did this affect the closest approach of Hah to its sun?
5. How do you think you can get Hah to stay the same distance from its sun no matter where Hah is in its orbit? Try it.
6. What is the shape of this orbit? Where is Hah's sun?

PROBLEM SOLVING ACTIVITY

Make a "Moonometer"

Use after page 132.

Background

An optical illusion is an incorrect image that is viewed by our eyes. Many times we look at something and our eyes play tricks on us. When we look at some objects, we see images that are not correct or exact. Every person with normal eyesight can experience an optical illusion.

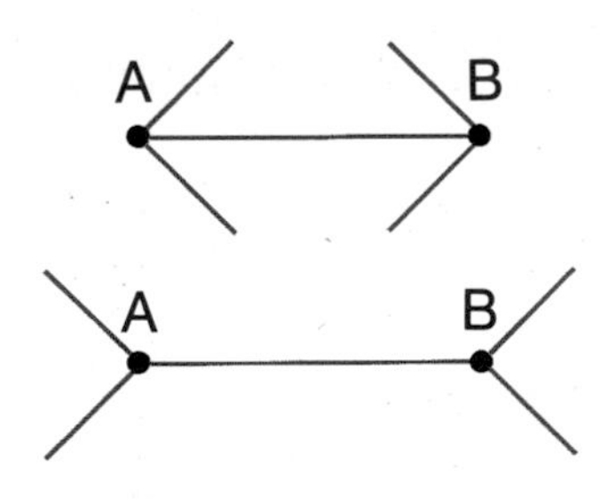

Which is longer?
Measure the lines.

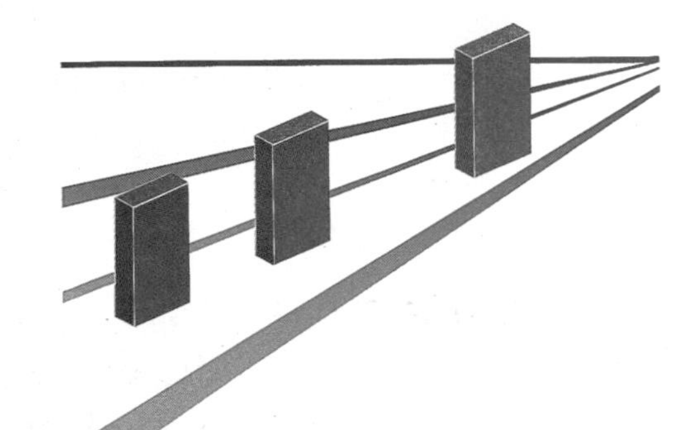

Which block is biggest?
Measure the blocks.

Problem

You have seen the full moon many times. Sometimes it is over the horizon. Sometimes it is high in the sky. Have you ever noticed that the moon usually looks bigger when it is over the horizon than when it is higher in the sky? Is the moon really bigger or is this an optical illusion?

Your problem is to think of a method of measuring the size of the moon as you see it in the sky. You will need to measure the moon twice; once when it is over the horizon and once when it is higher in the sky. Your "moonometer" should help you find out if the apparent size of the moon really is bigger over the horizon or if it is an optical illusion.

Materials cardboard • scissors • coins • other materials of your choice

Solution

1. Describe how you will make your moonometer.
2. Draw a picture of your moonometer.
3. Explain how your moonometer will work.
4. Record the results of your observations. What can you conclude about the size of the moon?
5. Explain the difference between the apparent size of the moon and its actual size.

PROBLEM SOLVING ACTIVITY

Battling an Oil Spill

Use after page 202.

Background When oil accidentally spills into the ocean, living things can be harmed. Sea birds can become covered with oil and drown. Shellfish can be poisoned. Fish used as food by people can have their oxygen cut off. For these and other reasons, it is very important to remove the spilled oil from the ocean as quickly as possible.

Problem Off the coast of California, two oil tankers collide. Thousands of liters of oil spill into the ocean. You are called upon to find a way to keep the oil from spreading and remove it from the water.

Materials large aluminum pan • motor oil • dropper • cotton balls • straw (hay) • detergent • watch with seconds indicator • metric ruler • pieces of: nylon net, nylon stocking, Styrofoam, cardboard, string, other materials of your choice

1. Fill the pan with water to a depth of 3 cm.
2. Use the dropper to slowly put 40 drops of oil on the water. Do this in the center of the pan.
3. Select any of the materials listed above to construct a barrier that will keep the oil from spreading. Describe the barrier.
4. Put your barrier where you want it. Have your partner use a piece of cardboard to fan air across the surface of the water. What happens to the oil?
5. If your barrier failed to keep the oil from spreading, write what you think went wrong. Change the barrier to make it better. Test it.
6. Repeat steps 1 and 2. Measure the diameter of the oil pool that forms. Write your measurements.
7. Select any of the materials listed above to remove the oil from the water. How do you plan to use the materials to remove the oil?
8. Make a table like the one shown below. Try your method for one minute. Measure the diameter of the oil pool that remains. Record the measurement in your Data Table.
9. Repeat step 7 with other materials. Circle the material in your Data Table that most reduced the size of the oil pool. Why do you think this material worked best?

Data Table

Material	How Used	Diameter of Oil Pool
none	control setup	

PROBLEM SOLVING ACTIVITY

What Is the Matter?

Use after page 218.

Background

In your travels, you find that you have mistakenly landed in a hostile space colony headed by the Keeper of the Light. She is assisted by three leaders—General Translucent, General Transparent, and General Opaque. Since you are thought to be a spy from another galaxy, you are quickly imprisoned along with other galactic offenders.

While you are desperately trying to think of an escape plan, the three generals begin fighting among themselves. Each of them thinks that his or her way is the only way that matter can affect light.

General Translucent feels that matter lets light pass through but scatters it. So, someone cannot see through it clearly. General Transparent argues that matter lets light pass through, letting someone see through it clearly. General Opaque insists that matter blocks light completely, so that someone cannot see through it at all.

The Keeper of the Light is tired of all the fighting among the three leaders. She seeks a solution to the constant arguing. She offers freedom to any and all prisoners who can solve the problem.

Problem The Keeper of the Light gives all of the prisoners a word puzzle that involves different examples of matter. She offers the reward of freedom to the prisoners who can find all of the words and figure out if any of the generals are telling the truth.

There is one additional problem. The Keeper of the Light has given the list of word clues in mirror writing. You must first decipher the list.

Materials small mirror

Solution

1. Copy the puzzle below. Then circle the words you can find from the list. They may appear horizontally, vertically, or diagonally.

WORD LIST

bricks
classroom floor
concrete walls
eyeglasses
frosted glass
kitchen wall
lamp shade
stained glass
sunglasses
waxed paper
windows
wood

s t a i n e d g l a s s e w m o s h e k n o s o r
n m u l o y n e h t e c l a p b s f l i t i v w l
t s i l d e r o t c e c b a t s p r e t t s x a b
l i c n e g p r e n l y w e m t r o d c h e d x e
f r o l p l d b s u n g l a s s e s r h e n u e p
w u l c l a s s r o o m f l o o r t t e w r o d e
a i l t s s m i t i l o r a c e d e k n o o w p e
h o n i t s e n v e c k s p u i l d n w g r i a n
b o o d r e r e s o r k n o o f w g o a o a s p o
s c i s o s k n l a m p s h a d e l e l o o n e r
t p y e e w b t c o n c r e t e w a l l s f d r i
l p a t e s s m a g f a r l y w e s h o o v n d r
n i e n l d r s t e d t c r a o l s m i t g e e d

2. Write the headings given below. Then write the words you circled in question 1 under the correct headings.

General Translucent	General Transparent	General Opaque

3. Are any of the generals telling the truth? Explain.

PROBLEM SOLVING ACTIVITY

Everybody Rock and Roll

Use after page 229.

Background

Vibration, volume, pitch, and frequency are all elements of sound. Music is based on a scale that combines these elements. A scale is a set of tones arranged according to pitch. These tones are known as the notes of a scale. The first seven notes of a scale are indicated by the letters A, B, C, D, E, F, and G. These letters are repeated every eight notes. These eight notes make up an octave. The notes below represent a scale in C major.

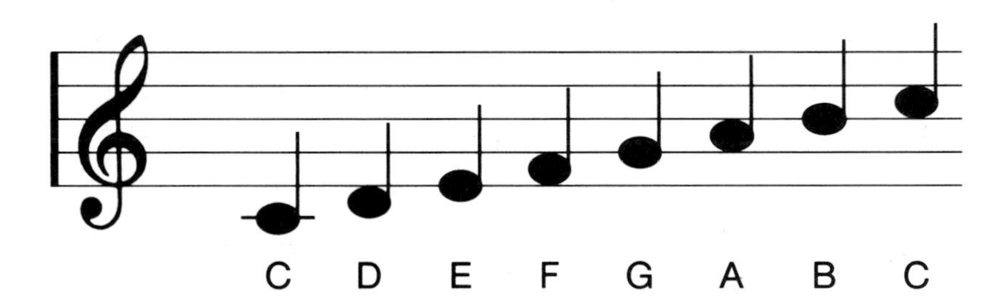

Singers often practice the notes of an octave using the syllables *do, re, mi, fa, so, la, ti, do.*

Problem

You and some of your friends want to form a music band. You have one set of drums, but no other instruments. You would like to have a variety of instruments in your band. Some of these instruments could include a guitar, keyboard, trombone, xylophone, marimba, harpsichord, or chimes. You and your friends have decided to make your own instruments.

Your problem is to construct an instrument of your choice. The instrument must have a full range of notes that can play the eight notes of a scale.

Materials

(suggested, but not limited to) tuning fork • boxes • rubber bands • fishing line • tin cans • wash basin • drinking straws • metal tubes • bottles • strips of wood • materials of your choice

Solution

1. Name the instrument you will construct. Describe the instrument.
2. List the materials you will use to make your instrument.
3. Describe how you will construct the instrument.
4. Draw a picture. Label the parts.
5. How can you test your instrument to see if you can play a full range of eight notes?
6. Explain how your instrument uses the properties of sound.
7. Try composing a song. Use the full range of octave notes.

PROBLEM SOLVING ACTIVITY

Magnetic Power

Use after page 258.

Background

Magnets and magnetism are used in our everyday life in many ways. Magnets are used in televisions, telephones, and radios to change electrical impulses into sounds. Navigators use compasses made with magnets to guide their ships. Large, powerful electromagnets are used in industry to lift heavy objects made of iron or steel. Electricity can be produced using magnetism. There are many ways that magnets are used in appliances in our homes, in industry, and in scientific research.

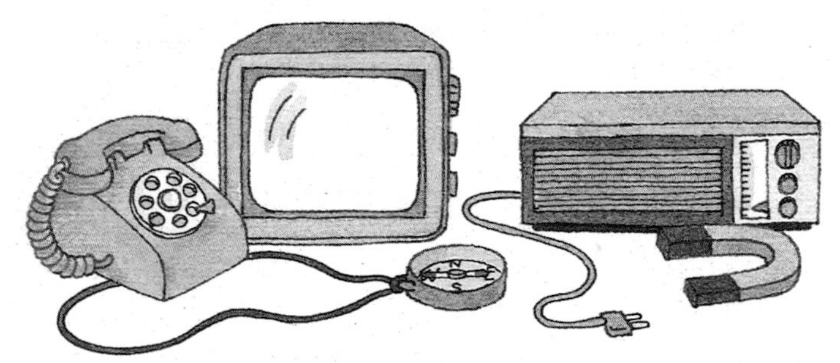

Problem

Your class has been brainstorming about the many ways that magnets are used in everyday life. You have decided to try to use a magnet in a new way. Your teacher has challenged you to build a toy boat that is operated by the power of a magnet. You must follow several rules in the construction of your magnet-powered boat.

First, each group of students must use the same type of magnet. Second, the magnet may not touch the boat. Third, you must set up an obstacle course to test your magnet-powered boat. You may use any materials of your choice to build your boat.

Materials magnets • large flat plastic basin • water • materials of your choice

Solution

1. Decide on the type of magnet you and the other groups will use.
2. List the materials you will use to build your boat.
3. Describe how you will construct your boat. Tell how your boat will make use of magnetic power.
4. Draw a picture of your boat. Show how you will use magnets.
5. Describe how you will construct an obstacle course to test each magnet-powered boat. Each group should use the same course.
6. Build and test your boat. Decide with the class how you will judge the most effective magnet-powered boat. Describe the results of your testing.
7. Which boat do you think is the most effective? Explain why you think so.

PROBLEM SOLVING ACTIVITY

Shop Smart!

Use after page 313.

Background

It is important to be a smart shopper. One reason for smart shopping is making sure you are getting your money's worth. Another reason is quality. Testing new products from a scientific viewpoint can help you determine the quality of a product and if you are getting the best buy.

Rating products is a good way to scientifically determine quality. When you rate a product, you must consider certain things about the product that you feel are important. When rating food products, you may want to consider things such as nutritional value, taste, and price. Other things, such as color, texture, and smell, may also be considered.

Problem

Your class has volunteered to investigate several food products. You are going to test and evaluate different brands of a certain product to determine the best buy. Your tests will depend upon the characteristics of the product that you decide are the most important.

For example, if you decide to investigate tomato juice, you may want to consider color, texture, taste, nutritional value, and price. When testing nutritional value, you may only want to consider one aspect, such as the amount of vitamin A. You must also devise a rating system to test each product in a scientific manner.

Materials products of your choice to be tested

Solution

1. Choose the food product that your group will investigate.
2. Decide, with your group, the five most important characteristics of that product. List them.
3. Devise a method of rating each characteristic. Describe your rating method.
4. Devise a method for testing each characteristic. Make a table like the one below.

Product Survey

Name of Product ______		
Characteristic	Test Method	Rating

5. Select two or three brands of your product to test. Collect and record the data you obtained by testing. Prepare a product report. What do you conclude from your investigation? Which brand would you buy? Explain why.

Glossary

This book has words you may not have read before. Many of these words are science words. Some science words may be hard for you to read. You will find the science words in **bold print.** These words may appear two ways. The first way shows how the word is spelled. The second way shows how the word sounds. The list below shows the sound each letter or group of letters makes.

Look at the word **Calorie** (KAL uh ree). The second spelling shows the letters "ee." Find these letters in the list. The "ee" has the sound of "ea" in the word "leaf." Any time you see "ee," you know what sound to say.

The capitalized syllable is the accented syllable.

a . . . back (BAK)
er . . . care, fair (KER, FER)
ay . . . day (DAY)
ah . . . father (FAHTH ur)
ar . . . car (KAR)
ow . . . flower, loud (FLOW ur, LOWD)
e . . . less (LES)
ee . . . leaf (LEEF)
ih . . . trip (TRIHP)
i (*or* i + *consonant* + e) . . . idea, life (i DEE uh, LIFE)
oh . . . go (GOH)
aw . . . soft (SAWFT)
or . . . orbit (OR but)
oy . . . coin (KOYN)
oo . . . foot (FOOT)
yoo . . . pure (PYOOR)
ew . . . food (FEWD)
yew . . . few (FYEW)
uh (*or* u + *consonant*) . . . comma, mother (KAHM uh, MUTH ur)
sh . . . shelf (SHELF)
ch . . . nature (NAY chur)
g . . . gift (GIHFT)
j . . . gem, edge (JEM, EJ)
ing . . . sing (SING)
zh . . . vision (VIHZH un)
k . . . cake (KAYK)
s . . . seed, cent (SEED, SENT)
z . . . zone, raise (ZOHN, RAYZ)

A

alcohol (AL kuh hawl): a legal drug for adults that is found in beer, wine, whiskey, and some other beverages

amphibians: cold-blooded animals that live part of their lives in water and part on land

arthropod: an animal that has an outer skeleton, jointed legs, and a body divided into sections

asteroids (AS tuh roydz): rocklike space objects that orbit the sun between Mars and Jupiter

atom: the smallest part of matter

B

battery: something that can produce an electric current

benthos (BEN thas): plants and animals that live on the ocean bottom

birds: warm-blooded animals that have feathers and lay eggs

black holes: stars that have gravity so strong that light cannot escape

C

caffeine (ka FEEN): a legal drug that is found in coffee, tea, chocolate, and some cola beverages

Calorie (KAL uh ree): a measure of the amount of energy in foods

carbohydrates (kar boh HI drayts): nutrients that provide the main source of quick energy for the body

cartilage (KART uh lihj): firm, flexible material that covers the ends of bones

chlorophyll (KLOR uh fihl): the green matter in plants that traps energy from the sun

circuit (SUR kut): the path through which a current flows

cleanliness (KLEN lee nus): the habit of being well groomed

cocaine (koh KAYN): an illegal and harmful drug made from the leaves of the coca bush

cold-blooded animal: an animal whose body temperature changes with the temperature of its environment

combination group: foods that contain foods from more than one group such as stew and pizza

comets (KAHM uts): space objects made of ice mixed with dust particles

conclusion (kun KLEW zhun): an answer to a problem or question

conductor (kun DUK tur): matter through which electrons can flow easily

continental shelf: shallow, sloping area beyond the surf

continental slope: area that begins where the continental shelf drops off steeply to the plain

corona (koh ROH nuh): the outermost layer of gases surrounding the sun

crack: an illegal and harmful drug made from a mixture of cocaine and other substances

crest: the high point of a wave

current (KUR unt): the horizontal flow of ocean water; the movement of electrons along a path

D

data (DAYT uh): recorded facts or measurements

diatoms: yellow-green algae that live in the ocean

diet (DI ut): all the food a person eats

dinosaurs: reptiles that died out millions of years ago

drug: a chemical that changes the way a person feels, thinks, or acts

E

echo: a reflected sound

electromagnet (ih lek troh MAG nut): a temporary magnet made using electric current

electron (ih LEK trahn): a particle found outside the nucleus of an atom

ellipse (ih LIPHS): an oval shape

embryo (EM bree oh): an underdeveloped plant or animal in its early stages of growth

exercise: any activity that uses the muscles of the body

F

fats: nutrients that provide the body with the most concentrated form of energy

fish: cold-blooded animals that live in water and breathe through gills

flatworms: simple worms with long, flat bodies and only one opening

frequency (FREE kwun see): the number of times an object vibrates in one second

fruit: a plant part in which seeds grow

fruit-vegetable group: foods such as apples, bananas, broccoli, and lettuce

fusion (FYEW zhun): the joining together of hydrogen atoms to form helium

G

generator (JEN uh rayt ur): a machine that produces an electric current

germination (jur muh NAY shun): the beginning of the growth of a plant embryo

grain group: foods such as bread, cereal, rice, and pasta

gravity: a force that causes objects to be attracted to one another

growth hormones: chemicals your body makes that cause you to grow

H

hypothesis (hi PAHTH uh sus): a statement that suggests or predicts an answer to a question

I

illegal drugs: harmful drugs that a person is not allowed by law to buy, sell, or use

insects: arthropods that have three body sections and six legs

insulator (IHN suh layt ur): matter through which electrons do not flow easily

invertebrates (ihn VURT uh brayts): animals that do not have backbones

inner planets: the planets closest to the sun: Mercury, Venus, Earth, and Mars

K

kelp: brown seaweed

L

leaves: the main plant parts in which food is made

legal drug: a drug that a person is allowed by law to use

limestone: rock formed from many layers of tiny animal skeletons and shells

lodestones (LOHD stohnz): natural magnets

M

magnet: an object that is able to attract some materials

magnetic field: the area around a magnet where the magnetic force acts

mammal: warm-blooded animal that is covered with hair or fur and produces milk to feed its young

marijuana (mar uh WAHN uh): an illegal and harmful drug made from the hemp plant

marsupial (mar SEW pee ul): a mammal with a special pouch for carrying the undeveloped young

meat group: foods such as hamburger, fish, chicken, eggs, and peanut butter

metamorphosis (met uh MOR fuh sus): the change in animals from young to adult through different stages

meteoroids (MEET ee uh roydz): pieces of metal or rock that orbit the sun

milk group: foods such as milk, yogurt, cheese, and ice cream

minerals (MIHN uh rulz): nutrients needed in every part of the body for mental and physical health

mollusks: soft-bodied invertebrates that live on land or in water and usually have a shell

moon: Earth's natural satellite

moon phases (FAYZ uz): the changing appearance of the moon as seen from Earth

N

natural resource: a valuable material supplied by nature

nekton (NEK tun): free-swimming animals that live in the ocean

neutron (NEW trahn): a particle in the nucleus of an atom

neutron stars (NEW trahn): collapsing supernovas

nicotine (NIHK uh teen): a legal drug for adults that is found in tobacco

nodules (NAHJ ewlz): mineral lumps found on the ocean bottom

nucleus (NEW klee us): the core of the atom

nutrients (NEW tree unts): materials needed by living things for growth; materials in food that are used by the body

O

oceans: large bodies of salt water

ocean waves: the up and down motion of surface water

oil spills: the release of oil wells or ships that can kill animals in the ocean or on the shore

opaque (oh PAYK) **matter:** matter through which light cannot pass

orbits: the paths objects follow when they revolve around a larger object

outer planets: the planets farther from the sun: Jupiter, Saturn, Uranus, and Pluto

over-the-counter drugs: legal drugs than can be purchased without a doctor's order

P

parasite (PER uh site): an organism that feeds on and causes harm to another organism

photosynthesis (foht uh SIHN thuh sus): the process by which plants make food

pistil (PIHS tul): the female part of a flower that contains the ovary

pitch: the highness or lowness of a sound

plain: the bottom of the deep, open ocean

planet: a large space object that moves around the sun

plankton (PLANG tun): very small organisms that drift with the ocean current

plant life cycle (SI kul): the process of germination, growth of a plant, and formation of new seeds

plaque (PLAK): a sticky material that forms on teeth and is harmful to dental health

poles: the ends of a magnet

pollen (PAHL un): flower part that contains sperm cells

pollination (pahl uh NAY shun): the transfer of pollen grains to the sticky part of the pistil

prescription (prih SKRIHP shun) **drug:** a legal drug that can be ordered only by a doctor

prism (PRIHZ um): a transparent object that refracts light

proteins (PROH teenz): nutrients needed for growth and repair of body cells

proton (PROH tahn): a particle in the nucleus of an atom

R

red giant: a large star that collapses

reflection (rih FLEK shun): the bounding back of light from a surface

refraction (rih FRAK shun): the bending of light

regeneration (rih jen uh RAY shun): the regrowth of cells or body parts of an organism

reptile: cold-blooded animal that has dry, scaly skin

rift zones: systems of cracks in the ocean floor through which magma rises

roots: the parts of a plant that hold the plant in the ground and take in nutrients

roundworms: simple worms with long, round bodies and two openings

S

satellite (SAT uh lite): any object that orbits or revolves around another larger object

scavengers (SKAV un jurz): animals that live on dead or rotting animals

scientific method (si un TIHF ihk •

METH ud): an organized way of asking questions, gathering information, and finding answers

seas: small bodies of salt water

seed plants: plants that form seeds

segmented worms: worms that have long, round bodies divided into segments

sleep: a state of restfulness

smokeless tobacco: tobacco that is either chewed or placed between the cheek and gums

solar eclipse (SOH lur • ih KLIHPS): an event during which the moon comes directly between Earth and the sun

solar flares: giant bursts of fiery gases that shoot from the sun's surface

solar system: the sun and all the space objects traveling around it

sonar (SOH nar): an instrument that uses sound waves to locate objects in the water

space probes (PROHBZ): spacecraft sent beyond Earth to gather information about space objects

sphere (SFIHR): an object shaped like a ball

spiny-bodied animals: animals that have five-part bodies, spines, and skeletons of plates

sponge: the simplest kind of animal without a backbone

stamens (STAY munz): flower parts that produce pollen grains

static electricity (STAT ihk • ih lek TRIHS ut ee): the charge of an object that has an unequal number of protons and electrons

stems: the parts of a plant that hold up the leaves and carry nutrients from the roots to the rest of the plant

stinging-celled animals: simple invertebrates that have stinging tentacles around their mouths

sun: the star closest to Earth

sunspots: dark spots that appear on the surface of the sun

supernova: an exploding red giant

switch: a device used to open or close a circuit

T

THC: a harmful substance in marijuana that stays in the body a long time

tide: the rise and fall of ocean water levels caused by the gravitational pull of the moon

translucent (trans LEW sunt) **matter:** matter that scatters light in many directions

transparent (trans PER unt) **matter:** matter that light can pass straight through

trenches: deep, narrow valleys in the ocean floor

trough (TRAWF): the low point of a wave

tsunamis (soo NAHM eez): huge ocean waves caused by earthquakes on the ocean floor

turbines (TUR bunz): devices that have blades attached to an axle; turbines provide power to run generators

V

vertebrates (VURT uh brayts): animals that have backbones

vibrations (vi BRAY shunz): the back and forth movements of particles of matter

visible spectrum (VIHZ uh bul • SPEK trum): the band of colors that make up white light

vitamins (VITE uh munz): nutrients that help the body use protein, fat, and carbohydrates

volume (VAHL yum): the loudness or softness of a sound

W

warm-blooded animal: an animal that can maintain a constant body temperature

white light: a mixture of many colors that can be separated into a band of colors

Index

E

F

G

H

I

J

K

L

M

N

O

P

R

S

T

U

V

W

Photo Credits

Cover: David Hall/Image Works
iv–v, Robert Bachand; **iv,** Brian Parker/Tom Stack & Associates; **v,** (t) Studiohio, (c) Brian Parker/Tom Stack & Associates, (b) Runk/Schoenberger from Grant Heilman; **vii,** Aaron Haupt/Merrill photo; **viii,** Elaine Shay; **xi,** Aaron Haupt; Merrill photo; **xii,** Carl Roessler/FPG; **xiii,** UP Photo/Bettman Archive; **xiv,** Kitt Peak Nat'l. Observatory; **xv,** Ken Reid/FPG; **xvi,** Doug Martin; **1,** (tl) Dan McCoy/Rainbow, (cl) Doug Martin, (bl) Kenneth Garrett/FPG, (r) E. Nagele/FPG; **2,** Kenji Kerins; **4,** H. Abernathy/H. Armstrong Roberts; **5,** Diane Graham-Henry and Kathleen Culbert-Aguilar; **6,** (l) Gary Milburn/Tom Stack & Associates, (r) Tom McHugh/Photo Researchers; **9, 10, 12, 13, 14,** Studiohio; **15,** (l) Ted Rice, (r) Studiohio; **17,** Ted Rice; **18,** (l) Animals Animals/Henry Ausloos, (r) Frank S. Balthis; **19,** Frank S. Balthis; **21,** William Weber; **22,** Ian Redmond; **23,** Peter Veit/DRK Photo; **27,** Mark Burnett/Merrill photo; **28,** Kenji Kerins; **29,** Animals Animals/C.C. Lockwood; **30,** (t) Carl Roessler/Tom Stack & Associates, (bl) Robert Bachand, (br) Animals Animals/Mickey Gibson; **31,** Milepost Corp.; **32,** Geri Murphy; **33,** Animals Animals/Breck P. Kent; **34, 35,** Milepost Corp.; **37,** Gary Milburn/Tom Stack & Associates; **38,** (l) Larry Lipsky/Tom Stack & Associates, (r) K.G. Vock, Okapia/Photo Researchers; **40–41,** Ruth Dixon; **40,** (l) Runk/Schoenberger from Grant Heilman, (c) Robert Bachand, (r) Susan Marquart; **41,** Brian Parker/Tom Stack & Associates; **43,** W. Marc Bernsau/The Image Works; **45,** Kirtley Perkins/Photo Researchers; **46,** Animals Animals/H. & J. Bester; **50,** Mack Albin; **52,** (l) Alvin Staffan, (r) Craig Kramer; **54,** (l) Paul Brown, (r) Alvin Staffan; **55,** (l) Alvin Staffan, (r) Roger K. Burnard; **56,** David M. Dennis; **58,** Frank S. Balthis; **59,** Alan Carey; **60,** Karl H. Maslowski; **61,** (l) Fritz Prenzel/TSW-CLICK/Chicago, (r) Dave Watts/Tom Stack & Associates; **63,** Milepost Corp.; **65,** Roger K. Burnard; **68,** Chris Rogers/The Stock Market; **69,** Luis Villotta/The Stock Market; **70,** Milepost Corp.; **72,** (l) Tom Pantages, (r) Roger K. Burnard; **73,** (l) Dwight R. Kuhn, (r) Grant Heilman Photography; **74,** Milepost Corp.; **75,** (l) Aaron Haupt/Merrill photo, (r) Milepost Corp.; **77,** (l) Runk/Schoenberger from Grant Heilman, (r) Dwight R. Kuhn; **78,** Ruth Bogart/Merrill photo; **79,** Gerard Photography; **81,** Tim Courlas; **84,** Kenji Kerins; **85,** Earth Scenes/Breck P. Kent; **86,** Studiohio; **87,** (tl) Roger K. Burnard, (tr) Studiohio, (bl) Frank S. Balthis, (br) Virginia Crowl; **89,** Tim Courlas; **90,** Elaine Shay; **94,** David M. Dennis; **96,** Studiohio; **98,** (l) Paul Brown, (r) Studiohio; **100,** Studiohio; **101,** NASA; **103,** Philip LaRocco; **104–105,** Aaron Haupt/Merrill photo; **106,** Carl Roessler; **108,** Gary Brettnacker/TSW-CLICK/Chicago; **109,** NASA; **115,** (t) Pictures Unlimited, (b) David L. Perry; **116, 117, 118,** NASA; **119,** (t) NASA, (c) Margaret Gowan/TSW-CLICK/Chicago, (b) file photo; **120, 122, 123, 124, 125, 126, 128, 129,** NASA; **130,** From "The World Book Encyclopedia", © 1986 World Book, Inc.; **131,** Studiohio; **132,** NASA; **133,** Tim Courlas; **135,** file photo; **136,** (l) Breck P. Kent, (r) John Stanford/Science Photo Library/Photo Researchers; **140,** NASA; **142,** Peter Langome'/Uniphoto; **143,** Diane Graham-Henry and Kathleen Culbert-Augilar; **145,** Tracy Borland; **146,** Latent Image; **150,** (l) NOAA, (r) Steve Lissau; **151,** NASA; **152,** Tersch; **153,** D.K. Lynch/Hansen Planetarium; **156,** (l) file photo, (r) Brian Warner/Tersch; **157,** Kitt Peak Nat'l. Observatory; **159,** Cobalt Productions; **162,** Dan Rest; **163,** E. Newrath/Stock Boston; **165,** Studiohio; **166,** Susan Marquart; **167,** (t) Larry Hamill, (b) Grant Heilman Photography; **169,** (l) file photo, (r) Joey Jacques; **170,** Cavataio-All Sport-Jim Richards/Photo Researchers; **173,** Porterfield-Chickering/Photo Researchers; **174,** O. Brown, R. Evans & J. Brown/University of Miami/RSMAS; **177,** William E. Townsend, Jr./Photo Researchers; **178,** Aramco; **180,** Jim McNee/Tom Stack & Associates; **181,** Fred McConnaughey/Photo Researchers; **183,** Rex Educational Resources Co.; **185,** Kees Van Den Berg/Photo Researchers; **186,** Howard Sochurek/Medichrome; **188,** Jan Robert Factor/Photo Researchers; **189,** Tony Roberts/Photo Researchers; **190,** Ulrike Welsch/Photo Researchers; **194,** Bernard Asset-Agencie Vandystadt/Photo Researchers; **195,** Manfred Kage/Peter Arnold, Inc.; **198,** (l, c) Studiohio, (r) Porterfield-Chickering/Photo Researchers; **199,** Stu-

diohio; **200,** (l) Larry Lefever from Grant Heilman, (r) Ron Levy; **201,** Jose Dupont-Explorer/Photo Researchers; **202,** Ted Rice; **203,** Studiohio; **205,** Geri Murphy; **206–207,** Aaron Haupt/Merrill photo; **208,** Tom Ulrich/TSW-CLICK/Chicago; **210,** Bob Daemmrich; **211,** Diane Graham-Henry and Kathleen Culbert-Augilar; **212,** (l) Bob Daemmrich, (c) Latent Image, (r) Lawrence Migdale/Photo Researchers; **213,** Gary Milburn/Tom Stack & Associates; **216,** Pictures Unlimited; **217,** Elaine Shay; **218,** Studiohio; **220,** Pictures Unlimited; **221,** (t) Roger K. Burnard, (b) Hickson-Bender Photography; **222,** (t) Studiohio, (b) R.D. Estes/Photo Researchers; **223,** (l) Ted Rice, (r) Doug Martin; **224,** Cobalt Productions; **225,** (l) Bob Daemmrich, (r) Pictures Unlimited; **227,** Janet Adams; **228,** Philip LaRocco; **229,** (l) Mary Lou Uttermohlen, (tr, cr) Studiohio, (br) Roger K. Burnard; **231,** courtesy of Eye-Dentify, Inc.; **233,** Mindy E. Klarman/Photo Researchers; **234,** COSI/Ohio's Center of Science & Industry; **235,** Rich Forman; **236,** Jack Sekowski; **238,** Mary Lou Uttermohlen; **240,** Gary Ladd; **241,** Aaron Haupt/Merrill photo; **242,** Brent Turner/BLT Productions; **243,** Aaron Haupt/Merrill photo; **244,** Studiohio; **245,** (tl, tr) Doug Martin, (b) Aaron Haupt/Merrill photo; **246,** (tl, tr) Doug Martin, (b) Studiohio; **247,** Brent Turner/BLT Productions; **249,** Ted Rice; **251,** courtesy General Motors/Hughes Aircraft/AeroVironment; **252, 253,** Diane Graham-Henry and Kathleen Culbert-Aguilar; **254,** Aaron Haupt/Merrill photo; **255, 256,** Brent Turner/BLT Productions; **257,** Latent Image; **260,** Aaron Haupt/Merrill photo; **261,** Studiohio; **262,** Aaron Haupt/Merrill photo; **263,** Brent Turner/BLT Productions; **265,** Lowell Georgia/Photo Researchers; **267,** Aaron Haupt/Merrill photo; **270–271,** Aaron Haupt/Merrill photo; **272,** Kenji Kerins; **274,** Bob Daemmrich; **275,** Ron Blakeley/Uniphoto; **276,** (tl) Studiohio, (bl) Studiohio, courtesy "The Wright Place" Inc., (r) Brent Tuner/BLT Productions; **277,** Steven Sutton/Duomo; **278,** (l) Studiohio, (r) Frank Lerner; **282,** H.M. DeCruyenaere; **284,** Brent Turner/BLT Productions; **286,** Lawrence Migdale; **288,** Brent Turner/BLT Productions; **289,** Mary Lou Uttermohlen; **290,** (t) Doug Martin, (b) Tim Courlas; **291,** Mary Lou Uttermohlen; **293,** Jack Sekowski; **294,** Doug Martin; **295,** Aaron Haupt/Merrill photo; **297,** Gerard Photography; **299,** Studiohio; **300, 301,** Bob Daemmrich; **304,** Brent Turner/BLT Productions; **305,** Studiohio; **306,** Gerard Photography; **307,** Studiohio; **308,** Hickson-Bender Photography; **309,** Studiohio; **310,** Elaine Shay, (inset) Brent Turner/BLT Productions; **311,** Nat'l. Dairy Council; **313,** Bob Daemmrich; **315,** courtesy of Dr. Richard J. Simonsen; **317,** (t) Mark Burnett/Merrill photo, (b) Brent Turner/BLT Productions; **318,** David York/Medichrome; **319,** Diane Graham-Henry and Kathleen Culbert-Aguilar; **320,** Latent Images; **321,** Bob Daemmrich; **322,** Shay/Gerard Photography; **323,** Jack Sekowski; **324,** Tim Courlas; **325,** Mark Burnett/Merrill photo; **326,** (l) Studiohio, (r) Aaron Haupt/Merrill photo; **327,** Larry Lefever from Grant Heilman; **328,** Aaron Haupt/Merrill photo; **330,** (l) © Puffons, 1983, (r) Image Workshop; **332,** Campaign/NBC; **333,** Ed Gallucci; **334,** Studiohio; **335,** Ron Nelson/Black Star; **338–339,** Aaron Haupt/Merrill photo.